ZOYA'S STORY

Zoya *with* John Follain
and Rita Cristofari

ZOYA'S STORY

An Afghan Woman's
Struggle for Freedom

wm WILLIAM MORROW
An Imprint of HarperCollins*Publishers*

"I'll Never Return" by Meena. Reprinted by permission of the Revolutionary Association of the Women of Afghanistan.

ZOYA'S STORY. Copyright © 2002 by John Follain and Rita Cristofari. All rights reserved. Printed in the United States of America. No part of this book may be used or reproduced in any manner whatsoever without written permission except in the case of brief quotations embodied in critical articles or reviews. For information address HarperCollins Publishers Inc., 10 East 53rd Street, New York, NY 10022.

HarperCollins books may be purchased for educational, business, or sales promotional use. For information please write: Special Markets Department, HarperCollins Publishers Inc., 10 East 53rd Street, New York, NY 10022.

Designed by Judith Abbate

Library of Congress Cataloging-in-Publication Data has been applied for.

ISBN 0-06-009783-3

09 10 RRD 10 9 8 7 6 5 4

To the women of Afghanistan,
victims of inhuman suffering
inflicted by fundamentalism

Contents

ZOYA'S STORY

Prologue

AT THE HEAD of the Khyber Pass, when we reached the border with Afghanistan at Torkham, our car stopped short of the Taliban checkpoint. Before getting out of the car, my friend Abida helped me to put the *burqa* on top of my shirt and trousers and adjusted the fabric until it covered me completely. I felt as if someone had wrapped me in a bag. As best I could in the small mountain of cheap blue polyester, I swung my legs out of the car and got out.

The checkpoint was a hundred yards away, and I stared for a moment at my homeland beyond it. I had been living in exile in Pakistan for five years, and this was my first journey back to Afghanistan. I was looking at its dry and dusty mountains through the bars of a prison cell. The mesh of tiny holes in front of my eyes chafed against my eyelashes. I tried to look up at the sky, but the fabric rubbed against my eyes.

The *burqa* weighed on me like a shroud. I began to sweat in the June sunshine and the beads of moisture on my forehead stuck to the fabric. The little perfume—my small gesture of rebellion—that I had put on earlier at once evaporated. Until a few moments ago, I had breathed easily, instinctively, but now I suddenly felt short of air, as if someone had turned off my supply of oxygen.

I followed Javid, who would pretend to be our *mahram*, the male relative without whom the Taliban refused to allow any woman to leave her house, as he set out for the checkpoint. I could see nothing of the people at my side. I could not even see the road under my feet. I thought only of the Taliban edict that my entire body, even my feet and hands, must remain invisible under the *burqa* at all times. I had taken only a few short steps when I tripped and nearly fell down.

When I finally neared the checkpoint, I saw Javid go up to one of the Taliban guards, who was carrying his Kalashnikov rifle slung jauntily over his shoulder. He looked as wild as the Mujahideen, the soldiers who claimed to be fighting a "holy war," whom I had seen as a child: the crazed eyes, the dirty beard, the filthy clothes. I watched him reach to the back of his head, extract what must have been a louse, and squash it between two fingernails with a sharp crack. I remembered what Grandmother had told me about the Mujahideen: "If they come to my house, they won't even need to kill me. I'll die just from seeing their wild faces."

I heard the Taliban ask Javid where he was going, and Javid replied, "These women are with me. They are my daughters. We traveled to Pakistan for some treatment because I am sick, and now we are going back home to Kabul." No one asked me to show any papers. I had been told that for the Taliban, the *burqa* was the only passport they demanded of a woman.

If the Taliban had ordered us to open my bag, he would have found, tied up with string and crammed at the bottom under my few clothes, ten publications of the clandestine association I had joined, the Revolutionary Association of the Women of Afghanistan. They documented, with photographs that made my stomach churn no matter how many times I looked at them, the stonings to death, the public hangings, the amputations performed on men accused of theft, at which teenagers were given the job of displaying the severed limbs to the spectators, the torturing of victims who had fuel poured on them before being set alight, the mass graves the Taliban forces left in their wake.

These catalogs of the crimes perpetrated by the Taliban guard's regime had been compiled on the basis of reports from our members in Kabul. Once they had been smuggled to the city, they would be photocopied thousands of times and distributed to as many people as possible.

But the Taliban made no such request. Shuffling, stumbling, my dignity suffocated, I was allowed through the checkpoint into Afghanistan.

As women, we were not allowed to speak to the driver of a Toyota minibus caked in mud that was waiting to set out for

Kabul, so Javid went up to him and asked how much the journey would cost. Then Abida and I climbed in, sitting as far to the back as we could with the other women. We had to wait for a Taliban to jump into the minibus and check that there was nothing suspicious about any of the travelers before we could set off. For him, even a woman wearing white socks would have been suspicious. Under a ridiculous Taliban rule, no one could wear them because white was the color of their flag and they thought it offensive that it should be used to cover such a lowly part of the body as the feet.

The longer the drive lasted, the tighter the headband on the *burqa* seemed to become, and my head began to ache. The cloth stuck to my damp cheeks, and the hot air that I was breathing out was trapped under my nose. My seat was just above one of the wheels, and the lack of air, the oppressive heat, and the smell of gasoline mixed with the stench of sweat and the unwashed feet of the men in front of us made me feel worse and worse until I thought I would vomit. I felt as if my head would explode.

We had only one bottle of water between us. Every time I tried to lift the cloth and take a sip, I felt the water trickle down my chin and wet my clothes. I managed to take some aspirin that I had brought with me, but I didn't feel any better. I tried to fan myself with a piece of cardboard, but to do so I had to lift the fabric off my face with one hand and fan myself under the *burqa* with the other. I tried to rest my feet on the back of the seat in front of me so as to get some air around my legs. I struggled not to fall sideways as the minibus swung at speed around the hairpin bends, or to imagine what would happen if it toppled from a precipice into the valley below.

I tried to speak to Abida, but we had to be careful what we said, and every time I opened my mouth the sweat-drenched fabric would press against it like a mask. She let me rest my head on her shoulder, although she was as hot as I was.

It was during this journey that I truly came to understand what the *burqa* meant. As I stole glances at the women sitting around me, I realized that I no longer thought them backward, which I had as a child. These women were forced to wear the *burqa*. Otherwise they faced lashings, or beatings with chains. The Taliban required them to hide their identities as women, to make them feel so ashamed of their sex that they were afraid to show one inch of their bodies. The Taliban did not know the meaning of love: women for them were only a sexual instrument.

The mountains, waterfalls, deserts, poor villages, and wrecked Russian tanks that I saw through the *burqa* and the mud-splattered window made little impression on my mind. I could only think ahead to when my trip would end. For the six hours that the journey lasted, we women were never allowed out. The driver stopped only at prayer time, and only the men were allowed to get out of the minibus to pray at the roadside. Javid got out with them and prayed. All I could do was wait.

PART ONE

A Present *from* Russia

Chapter One

KABUL WAS ALWAYS more beautiful in the snow. Even the piles of rotting rubbish in my street, the only source of food for the scrawny chickens and goats that our neighbors kept outside their mud houses, looked beautiful to me after the snow had covered them in white during the long night.

I was four years old that December, and I had been playing in the snow with some other children. We pushed and shoved one another, trying to dodge the snowballs—not an easy thing to do in a street that was so narrow only three adults could walk down it shoulder to shoulder. We stopped playing because one of us wanted to buy something; not me, I didn't have the money, although the shopkeeper near my house usually let me pay for something the following day. We all piled into the small shop.

There was a Russian woman soldier in the shop when we entered. Like the soldiers I had seen marching in the city, she wore a dark green uniform and big boots. She saw me and stretched out her hand to offer me a chocolate in shiny yellow wrapping. It was one of my favorites.

The woman soldier towered over me and said something that I did not understand. It was the closest I had ever gotten to a Russian invader.

I had no idea what to do. I stared at her face. She looked just like the doll I had named Mujda (good news)—yellow hair, white skin, and green eyes. The kind of face that Grandmother had warned me about. "You should be scared of them," she would say sternly. "They are the invaders who have occupied Afghanistan. Their hands are stained with red, with the blood of our people. If an invader from Russia offers something to you, don't accept it, and don't go anywhere with them." But she had always talked about the men. She had never said anything about women.

The woman soldier came closer, thrusting the chocolate at me. I looked for the blood on her hand. I was afraid that if I touched it, my hand would have blood on it too. I thought that the blood would never come off me, however much I washed. But there wasn't any blood on her hand. I said no to her, but she just laughed. She said something to me, but I didn't understand.

She said something to the shopkeeper, and he spoke to me. "She says she likes you and she just wants you to accept this chocolate as a present from her. Why won't you accept it?"

I repeated what Grandmother had told me to say if a Russian ever spoke to me. "Well, if she is Russian, tell her to get out of my country." Then I walked out into the street.

But the woman soldier followed me. I stopped, and she just stood in front of me, and I could see that she was crying. She pulled a handkerchief out of her pocket and pressed it to her eyes.

I had never seen an invader cry before. I felt sorry for her. I would have liked to accept her present, but at the same time I was afraid of what Grandmother would think of me. I wanted to say, "Please wait. I will go ask Grandmother if I can have her permission to accept this chocolate or if I have to say something else to you." But the words stuck in my throat, and I scurried away home.

I jumped over the tiny smelly stream that ran past our house and that we all used as our sewer, pushed open the blue metal door with the flaking paint, and crossed the yard that I called our garden, even though flowers never grew there.

I kicked off my shoes and rushed across the brightly colored carpets to the spot where I knew I would find Grandmother. She spent almost all her day in a corner of the main room of our house, wearing a small veil over her hair and sitting on the floor on a *toshak*, a kind of mattress big enough for five people to sit on that was placed on top of the carpets. Sometimes she would lean against the wall of dried mud.

She was surrounded by her *taspeh* prayer beads, which she had in her hand all day long, the spray for her asthma, and the medicine she took for her rheumatism. No one else I knew prayed for as long as Grandmother. I had seen other people pray for two minutes and then get up again, but Grandmother would spend half an hour on the special prayer mat that she usually kept rolled up against the wall. I would be wanting something from her or to go out for a walk with her, but I would have to wait and wait until she finished.

The copy of the Koran, which she let me touch only after I had washed my hands, was also within easy reach on a small wooden table, protected by a cloth. She was weak and she had trouble getting up, so she did everything in the same place, from peeling vegetables to praying to Allah five times a day. When she did work in the kitchen, she moved so slowly that it was a long time before meals were ready.

But she was taking her early-afternoon nap on the mattress, and I didn't dare to wake her up because she had difficulty sleeping. I sat in front of Grandmother and tried to keep quiet. The minutes ticked by, so slowly. I picked up her brown beads and played with them for a while. When Grandmother prayed, she would mutter something under her breath, and the beads would go *click, click, click* as she ran them through her fingers. I had once asked her what she was saying, and she told me that she repeated my name, over and over again. I believed her and was happy that Grandmother would say my name all day long.

I kept thinking about the woman soldier. I felt ashamed as if I had done something wrong. When Grandmother woke up at last, I told her what had happened: "Grandmother, the woman was crying. I felt sorry for her. I refused because you always told me to refuse, but perhaps I should have accepted?"

"Daughter," Grandmother said—she always called me "daughter"—"it's not because she was a woman and she was crying that you should accept. You should have said thank you to her, but you did right to refuse the chocolate. Of course, not all Russians are bad. Some are like you and me. But never forget that they have entered your country without being invited and that they are forcing you to do what they want. They want you to be their servants. They want to steal the most precious things from our great mountains. But we want to decide our own future."

The Russians had been occupying Afghanistan for the past three years. They had invaded in December 1979, when I was one year old, with the excuse of bolstering a Marxist-Leninist government that had seized power in a bloody military coup because they feared that Moslem fundamentalists, backed by the Americans and the Chinese, would overrun the country as they had done in Iran, where they overthrew the Shah. The invasion dragged Afghanistan into the cold war, as the Mujahideen turned to America for help to fight the occupiers.

MY PARENTS' FAMILIES could not have been more different. Father was from a town in the south of Afghanistan. Like Mother, he was a member of the Pashtun tribe, the traditional rulers of Afghanistan, and spoke Persian. But Mother, whose parents had sent her to school and planned to send her to university as well, didn't think highly of his family. She thought they were backward, and they never came to our house.

"All the women in his family wear the veil," Mother told me. "They think it is quite normal to sell a girl into marriage for the price of a few cows or some sheep. Your father quarreled badly with his own father because he took two more wives when your father was still a child."

My parents were distant relatives, and their marriage was arranged by their parents, which was customary. But there was nothing customary about the way they celebrated the wedding. Usually the celebrations last a whole week. Even the poorest families will borrow huge sums of money to hold separate parties for the bride and the groom and offer meals to more than a thousand guests. Three hundred guests is considered too few. And each day the bride has to wear a new and expensive dress—each day a different color, as if she has to work her way through the rainbow.

Mother, who was eighteen years old when she got married, thought all this was ridiculous, just like the tradition in the more isolated villages that the couple must hang out the bedsheet after their first night so that people can tell from the blood stains that the bride was a virgin.

She insisted on a small celebration. She said the success of the marriage would not be measured by how much was spent on the wedding. She even decided not to go to the beauty salon on the day of her wedding. She joked that they would put so much makeup on her that she would be five kilos heavier.

The bride's ideas were too advanced for Father's relatives. Some of them said that small celebrations were held only when the bride was a widow or when there was something wrong with her—for example, if she was sick. But Father was the first in his family to get a proper education, having studied biology in Kabul, and he supported Mother. He liked simple things, and I can't remember him ever talking about money. My parents got their way, and there were forty guests at the wedding.

They were married in the house I grew up in, just four dark rooms with walls of hard mud—sometimes old, dry bits of mud would fall down on me from the ceiling as I played—and it was over very quickly. There's a tradition that the bride and groom must sit side by side and that the first they ever see of each other must be their reflections in a mirror that is held in front of them. But my parents' families arranged for them to meet a few weeks before they were married. They exchanged only a few words, and Father gave her a gold engagement ring.

There was no mirror at their wedding. Mother just sat next to Father. She wore a simple pale pink dress, and her only jewels were the ring that he had given her and some earrings. She didn't even use henna on her hands.

The mullah came in, dressed in clean white clothes, and asked the bride and then the groom whether they were willing to be married. The rule is that he asks each of them three times—even if they remain silent, it counts as a yes. Then they signed a piece of paper, and that was it. The celebrations lasted just one day. The guests got kabuli rice, which is our national dish—rice with chicken, cabbage, carrots, raisins, almonds, and pistachio nuts; *bolani,* which are fried potato cakes; and then dessert and fruit. According to Grandmother, Father's relatives said the wedding was like a funeral.

Although their wedding had been arranged, my parents grew to love each other. He respected her rights from the start. Many Afghan men believe that if their wives have been studying or working, they should drop everything and stay at home from the moment they get married. Father made no such condition, and Mother was free to continue literature studies at the university built by the Russians. She had rejected a couple of rich suitors who had much land and many horses because they were too traditionally minded. Father would never have dreamed of having more than one wife.

My parents rarely showed their affection in front of me. Mother liked to read love poems, and sometimes she would read them aloud to Father. They never let me see them kiss, but when Mother was tired late in the evenings she would ask Father to give her a massage. They would let me sit nearby as she stretched out on the bed, and he would start by massaging her head. Then he would massage her neck and shoulders as well, his hands over her nightshirt.

Grandmother told me that Father's relatives complained some more when I was born. Many of them were disappointed that I was a girl. I was my parents' first child, and for most Afghans it is important that a first child be a boy. In the traditional families in the villages, if the newborn is male, the family starts shouting, "A boy is born! A boy is born!" and the men fire guns into the air to celebrate. Relatives and friends bring money as a present and push it into the bed of the baby or the mother. In some families, boys would be given more to eat than their sisters.

When a girl is born, there is no shouting, and no one rushes to the house to congratulate the parents. And there isn't as much money in the bed. People go up to the mother and say, "Don't worry. Your next child will be a boy."

Father's family wanted a boy because he would grow up stronger than a girl, and when my mother grew old she would be able to go and live in her son's house—there was no such tradition with a daughter. But Father argued with his relatives and said he would love me just as much.

Father always told me, "When you grow up, you must become a doctor. Or you could become a good teacher, who would go out and educate the people." He had so many plans for me that Mother would laugh at him and say, "I have only one daughter, and you want her to do so many things. She can't realize all your dreams." Father would look serious and reply, "You'll see. She will do something good. She has what it takes."

When I asked Grandmother what she thought about me being a girl, she said to me, "I am happy about it. I feel that you are like both a son and a daughter to me, and I want you to be so strong in the future that no one will think that you are a woman."

I asked her how I was born.

"One day I was walking in the street," she replied, "and I saw this beautiful baby in a shop window. I stopped and stared at it, then I went into the shop, and I told the shopkeeper that I had no money, but please could he give me this beautiful daughter? But he said no, you were very expensive. So I went begging in the street, I got the money, and I bought you. That's how you were born."

I was proud that I had been an expensive baby.

Chapter Two

FROM AS EARLY as I can remember, I missed Mother. She was tall and slim, with big black eyes and beautiful black hair. When she had time, she would cook something special for me, like chicken and rice, or lamb, and then we would play hide-and-seek or blindman's buff. I knew she cheated and could see through the scarf over her eyes, but I didn't mind.

But most days she left the house early in the morning and came back late in the evening. She was so tired and was obviously working so hard that I did not dare ask her what she was up to. I knew that during the daytime she had something perhaps more important than me to take care of, but I also knew that every evening she would be at home and all mine. I resigned myself to losing her during the day.

Often I was told to say that Mother was not at home if someone asked for her at the door. She did not like to talk about her work in front of me. But I could see that some of the grown-ups in her family were unhappy about it, including Aunt Naseema, one of Mother's sisters. She was very different from Mother. I would hear her high heels clattering across the yard before I saw her, she wore a different dress every time, and she would often reapply her bright red lipstick. She liked luxury.

One summer, when I was five years old, I overheard Aunt Naseema tell Mother as they took tea together in our house, "Think about your husband and your daughter. You should get a proper job so that you can afford better things for yourselves and spend more time with your daughter. After all, she's the only one you've got."

"Nobody is forcing me to do this," Mother replied. "It's my decision."

"But think of the risks. The police and the secret service are so active these days, and you're going to all these dangerous places!" Aunt Naseema insisted.

Their voices had become so loud that I got up to leave the room. My parents never liked me to stay nearby when voices got loud. But I hadn't gone very far before Mother cut short the discussion: "I've heard enough. I don't want to waste my time. That's the only thing that matters to me."

I was angry with Aunt Naseema for speaking to Mother in a loud voice. She had no right to tell Mother off like that. I had no idea what Aunt Naseema's comment about the police and the secret service meant.

After Aunt Naseema had left, and as Mother was preparing to leave the house yet again, I found the courage to ask her, "Mother, why are you always so tired, and why are you always away from me? The other mothers in our street are at home with their children every day."

"I'm very sorry, Zoya, but I have a lot of work to do with my friends. I wish I could stay at home with you, but I have so many, many things to do," she replied.

I never asked my parents for a little brother or sister. But I overheard Grandmother telling my parents they should have another child. "There is nothing better than a big family, and that way your name will live on," she said.

I asked Mother why she did not want another child. She replied with a smile, "Even you are too much for me!" She never wanted to discuss this seriously. I now think that she realized she would not have the time to dedicate to another child. But I was glad that she didn't want anyone else—I would have been jealous and would have hated sharing her.

I understood that in spending little time with me, Mother was different from the other mothers in my street, but that was all I understood. I felt lonely, different from other children. The fathers of the other children went away to work every day, just as mine did, but the mothers stayed with them. Whenever I was sent to bed before Mother came back home, I couldn't get to sleep. Grandmother wasn't much help. She too had trouble getting to sleep, although she pretended she wasn't worried.

"Daughter, get to sleep now. I'll make sure Mother is fine when she gets back. Mother is doing important work, and one day you must try to be like her."

A FEW WEEKS after Aunt Naseema's visit, I was flying a kite with Father—one of the rare times that he had a day to spend with me—from the flat roof of our house and trying not to step into the tomatoes spread out in a corner to dry in the spring sunshine, when we heard a knock at the outside door. I was too small to be much good at flying kites. I would often lose my grip on the string, and the kite would slip away from my hand, its tail waving me good-bye as it soared up into the sky, high above the minarets and domes of the mosques, even above the mountains that surrounded my city and made me feel protected.

Father would never become angry with me. Instead he would go and make another kite with bright blue, red, and green paper that he had sent me to buy from a shop nearby. There were always many kites flying above Kabul, hundreds of them, more kites than birds. The kites would attack one another, and Father was a master at sending his slicing through the air to cut the string of another one, so that it too would climb and climb and then vanish.

He sent me to see who was waiting at the door that separated our little garden from the street, and I made my way slowly down the ladder, then across our dusty garden. I opened the door and was surprised to see a woman wearing a dirty yellow *burqa*. She was alone.

In the streets of Kabul, I had often seen women hiding underneath these *burqas*. They looked so strange next to the beautiful young women of the city who walked happily arm in arm, wearing makeup and short skirts. Afghan women had won the right not to wear the veil at a heavy price: when the prime minister and other ministers appeared in public in 1959 with their wives and daughters unveiled, the mullahs provoked riots that had to be crushed by the army.

Five years later, a new constitution proclaimed equality between men and women. But the religious leaders never accepted women's entry into the workforce, and when the puppet regime said girls should be free to choose their husbands and then in the late 1970s made female education compulsory, the fury of the mullahs, which was strongest in the rural areas, put the government's survival in jeopardy, and it turned to the Russians for help.

I would stare at the *burqas* and try to imagine what kinds of faces were under them. I wanted to ask the women why they were wearing them, but I never dared to. Never had one of them called at my house before. My parents had told me that only women from faraway villages who could not read or write wore them. The beggars wore them, and so did prostitutes who did not want to be recognized.

I didn't want to let the woman in. I stared hard at the grille in front of her face, but I could see only part of her eyes, her eyebrows, and the top of her nose. I couldn't even tell the color of her eyes. I didn't know who she was, and I wondered why she was hiding like this.

The woman spoke in a strange voice, as deep as a man's. "Zoya, my dear little Zoya, why don't you let me in?" she said.

It frightened me that this woman knew my name, and I wanted to close the door and run back to Father. But she stepped across the filthy stream and came into the garden. She closed the door behind her and suddenly bent down to lift off the *burqa* in one big sweep that blew a cloud of dust in my face and made my eyes sting. Then I saw that the woman was smiling at me.

It was Mother. I hugged her with relief, burrowing my nose into her neck to breathe her perfume, but I was upset that she had played a trick on me, and told her so. "Mother, why did you wear this?" I asked, pointing at the *burqa* draped over her arm. Whenever we had seen women wearing it in the streets, she had told me that she hated it and that it just made them more tired.

"Because it is so attractive," she answered with a laugh.

I knew she was joking, but she did not tell me the real reason. I quickly forgot about the *burqa* as Mother followed me back to the roof to join Father. I had both my parents all to myself. Mother was not as good as Father was at flying kites, but she enjoyed it too.

That is my first memory of seeing Mother wearing a *burqa*. It was only much later that I learned she had been wearing it to save her life. I could not imagine that one day I, and all the women not only of my city but of my entire country, would have to wear it for the very same reason.

THE EARLIEST MEMORY I have of the fighting between the Mujahideen and the Russians in my city is of a February night in 1984, when I was six years old. I remember waking in my bed with a start. I was used to waking up briefly when the cock crowed, and when the *azzan*, the preacher at the mosque, made his first call to prayer. Grandmother would get out of bed when she heard him, kneel down on her prayer mat, and start praying, and I would fall back to sleep.

But this was different. The only noise was that of the termites, their tiny jaws making a continuous scratching sound as they ate their way through my table and chair—they were so hungry that sometimes an arm of my chair would drop off while I was doing a drawing, and I had got so used to them I didn't mind sharing my bedroom with them. I didn't know what had woken me up, but it wasn't long before I found out. There was a sound of something heavy striking the ground with a thud. Then again. And again. I had no idea how close or how far away it was, but I did not want to stay on my own. So I threw off the blankets and ran barefoot in my nightshirt out of my room, across the garden where the cold air cut through me, and into Grandmother's room, where I climbed into her bed.

She was hard-of-hearing and was sleeping soundly. "What's happening?" she mumbled with a startled expression on her round face when I shook her awake.

"Just listen. What is it?" I said.

Then the sound came again. Grandmother pressed me to her chest. Her breasts were heavy and warm, still smelling of the talcum powder she bought at the bazaar, and her stomach was so big and soft it felt like dough. I liked the smell of her long hair, which was black and gray and which she usually wore in a bun. "All right, you can stay right here and sleep with me," she said.

"But what is it?" I insisted.

"It's nothing, daughter, don't worry. Go to sleep." And to send me to sleep, she told me a story in her soft, kind voice about a beautiful fairy and how the king wanted her all for himself.

The next morning my parents told me that what I had heard was the noise of soldiers practicing. But from then on, I almost always slept with Grandmother, either in her bed or in my own bed, which had been moved to her room. She told me to go to her at night and not to my parents, saying that I might disturb them and that she would always like to have me at her side. I was happy to obey her because I found more comfort in her presence than in that of my parents.

Despite what the adults told me, I started having a nightmare. It was always the same. I dreamed that I was on top of a mountain and that suddenly, without knowing whether I had jumped or been pushed by someone or something, I was about to fall through the air. I would wake up just before I started falling.

A few days later I heard the sound again. I was sitting in the main room drawing a birdcage like the ones I had seen in the bazaar, and Grandmother was in the kitchen—there was always the smell of something cooking in the house, rice or beans or vegetables of some kind—when zigzag cracks appeared across two windowpanes. The whole house trembled. I trembled too and dropped my colored pencil in fright.

For a moment I just stared, without moving, at the windows. I couldn't understand how two windows could have cracked at the same time—had someone played a trick and thrown two stones together just to frighten us? But I hadn't seen anyone in the garden.

Grandmother ran into the room and scooped me up in a soft hug. I didn't want to draw anymore after that. All I could think about was whether more windows would crack, and whether the glass would fall in on top of me.

Later she explained to me what a bomb was—that men could throw them from the mountains around my city, so far that they would explode in the middle of a house when everyone was sleeping there so that no one would ever wake up. From then on I always insisted on sleeping with the light on. Still today, I sleep with a light on.

Over the years of the Russian occupation, I was to hear that sound many more times whenever the Mujahideen came close to

Kabul and shelled the city's outskirts and, more rarely, the city itself. The Russians would manage to push them away for a time, but then the warlords would close in again, and the sound would return.

My sleep was easiest when, during the winter, my whole family slept in the same room, huddled around the *sandali*, which was the only source of heat in the house. Grandmother would put a plain lightbulb under a small table and then spread blankets over the table, and we would all sit cross-legged around the bulb under the blankets. The heat of the bulb was enough to keep us warm, but we had to be careful not to burn our feet. Because of the *sandali*—which Grandmother said made me even lazier than usual in the house—I always looked forward to the first day of the winter and to the hours we would spend drinking tea and talking around it.

I hated tearing myself away from it, even just to go brush my teeth and change into my nightclothes, because it was so cold in the rest of the house. Sometimes it would get so cold that the snow would harden into ice on the roof of our house, and Father would have to hack it off to stop the ceiling from falling in. When we slept by the *sandali*, we left the bulb on all night to keep us warm.

BECAUSE BOTH MY PARENTS were away so often, it was Grandmother who brought me up. She is the most important person in my life—more important even than my parents—

because she spent more time with me than they did. One of the reasons I love her so much is that, unlike my parents, she never forced me to sleep in the afternoons. I call her "Grandmother," but she isn't my real grandmother. My real grandmother died when I was very young.

Once I heard Aunt Naseema talking with Mother, saying that Grandmother was not my real grandmother. Mother was angry with her and told her to be quiet. I never asked how my parents had met Grandmother. I never wanted to know. And whenever Grandmother touched on the subject, I made her stop. "You are more real to me than a blood grandmother," I said to her.

Mother knew of my strong love for Grandmother, but she never did anything or said anything that made me think she was jealous. It was as if Mother wanted Grandmother to protect me.

Grandmother's name is Nabila. She was richer than my parents, and she had come to live in our house after the death of her husband, when I was seven months old. She left her three grown-up children to be with me, and she always told me that she loved me more than anything in the world.

When I was six years old, I began the ritual of rushing to her with a book I had sneaked out of Father's study when we were alone in the house. Father always told me to keep out of his study, but his books had a strong attraction for me—despite the fact that I didn't know how to read.

I liked the fact that all the books were different colors, and I knew that they were full of stories because often when I asked

Mother to tell me a story, she would go and fetch a book. She would open it for me, look at the book, and the story would begin.

I had to be careful. I didn't mind if I fell and hurt myself, but the worst thing would be if I damaged one of Father's books. I knew he would be furious if I dropped a book by mistake and the binding broke, or if I got a tea stain on one of the pages.

I would choose a particularly heavy one, which I thought would contain a specially good story, and bring it to Grandmother. She would settle down on the *toshak*, put on her glasses, and then she would start reading. The stories all began the same way: "Once upon a time there was a little girl who was alone with her grandmother. . . . "

At least I thought she was reading. I had no idea that the books I brought Grandmother were about biology, chemistry, or politics. All I knew was that there were no pictures in the books but that the stories were usually good. Now I know that Grandmother, although she knew how to read, did not under-stand a word of the books I brought her, and that she invented her stories as she went along. She had never gone to school. She didn't even know what year she was born in.

I always asked Grandmother to promise never to tell Father that I had taken the books from his study. Once, when I had dropped some tea on a page, she told Father that she had asked me to fetch her the book from the study.

Mother did not defend me. "You have your own books, in your own room. Why do you have to go and take Father's books?" she would say. But my books were dull because I already knew what was in them. In any case, as Mother was always out of the house or busy with the housework, and as Mother had less power inside the house than Grandmother, I didn't take any notice of her disapproval.

Grandmother's stories filled my days. Stories about the kings who used to rule Afghanistan, and how they tortured and killed their own people. She told me about the histories of the many different tribes that make up Afghanistan. They make it impossible for many historians to decide whether the country, which is Moslem but not Arab, belongs more to Central Asia, the Indian subcontinent, or the Middle East.

She told me about the Pashtun tribe to which I belong, insisting that although this tribe had given many rulers to Afghanistan there was nothing to be proud about. Several monarchs never fought for independence and gave in to whoever had invaded Afghanistan at the time. One king, Amir Abdul Rahman Khan, was famous for two things: his huge harem and his determination to wipe out the Hazara tribe, a tribe that Grandmother told me had Chinese-looking faces. This king built a tower with the heads of the enemies he had killed. Grandmother did not know how high the tower was.

Her stories made me laugh or taught me a lesson. In one story, the little girl is told never to open the door of the house to a Russian soldier—"you can recognize them easily because they have yellow hair, white skin, and green eyes"—but then the soldier forces his way into the house and tries to take the little girl away. She shouts for help and her family comes to rescue her.

She told me stories about the rich and the poor, and the ending was always the same: "Some people are rich, some people are poor, and the poor have to fight harder to survive and for their rights. You too will have to fight someday," she would say.

I liked her stories, but my attention drifted when she started talking about my future. She would stare at me and say in a stern voice, "Daughter, pay attention to what I am saying. Perhaps today my words have no meaning for you, but one day you will realize that they were important words."

I know now that I should have listened more closely, and that she was a good teacher to me. Even though she had not herself struggled against the traditions of her culture, unlike most of the women of her time she hoped that I, and other women of my generation, would learn from her mistakes.

Chapter Three

ONE MORNING a few days before my seventh birthday, Mother called me to her side as she sat in front of her dressing table. She lifted me up and sat me on her knee. I treasured such moments. I would ask her to let me have a few drops of her perfume, or at least to smell it on her from close up. I wanted to know where exactly the smell of perfume came from. On the table there was a cloth that Grandmother had embroidered with flowers, and on top of the cloth, alongside her toiletries, Mother kept her perfume. There were never more than two small bottles because it was so expensive. Mother always said that if perfume was cheap and no good, she would rather go without.

Perfume was the best present you could give Mother, and I once bought her some in the bazaar. It cannot have been one of her favorites, but she covered me in kisses all the same. She always said to me that the ones from Paris were the most famous in the world. Charlie was one of her favorites.

Usually Mother would not let me anywhere near her bottles. Many times she scolded me for putting some of her perfume on when I was all dirty after a day spent playing in the garden or in the street. "You're a bad girl. Have a bath and then use the perfume!" she would say.

But on this morning of 1985 she let me use a little, and then she looked serious and stared into my eyes. "Zoya," she said, "would you like to come with me when I leave the house to do my work?"

I felt so proud I said yes immediately, without thinking of asking what I should do. I found out soon enough. She asked me to fetch my backpack, one that had a picture of a bear on it. Then she packed some toys, my bottle—until the age of seven, I refused to drink milk unless it was in a bottle—and some papers I had never seen before, and sat me on her knee again to give me my instructions.

"If anyone stops us and asks where we are going, you must say that we are just going shopping. You must never say anything else, and you must never mention the papers in your knapsack. If someone finds them, you must say that you knew nothing about them. Will you remember all this?"

I nodded. Mother looked satisfied, and put on the horrid *burqa*.

"Why are you wearing this? I thought you hated it," I said.

34

"I do hate it, but I have to wear it because otherwise it is impossible for me to do my work," she answered.

We went hand in hand into the street. We went to different houses, and she would stop a few minutes at each of them. Sometimes she took me into the house with her, and I saw her talk hurriedly with people and give them one of the papers. Other times she left me in the street. "Look out for any soldiers, or for anyone you think might be a spy for the police," she would tell me.

Grandmother knew that I was running only a small risk given my age, and approved of my new job. Father did not want to interfere; he respected the work that Mother was doing. Mother seemed to be happy with my contribution, although sometimes I would forget that she had asked me to do something, and she would be angry with me.

It was strange and tiring work because Mother's trips sometimes lasted for hours. I didn't think my job in itself was very important, but I was only a small child and already I was helping Mother; that was all that mattered. And Mother had chosen me—I had heard Grandmother offer to help her, but Mother had turned her down.

MY OUTINGS with Mother were rare, and I looked forward to them with impatience. There was nothing I loved more than being with adults, whether it was Grandmother or my parents. I never enjoyed the company of other children. I thought their games were stupid, and I didn't like the way they would mock me about silly things the way children do.

I preferred staying in the house with Grandmother. Other children, my cousins, would ask Mother, "Why is she like this? Why does she not play with us?" This irritated Mother. "Be a child," she would say. "Don't sit here with Grandmother and me."

But I would lock my door when the children came looking for me shouting that the toy seller—who carried a stick from which he hung all sorts of balloons, little dolls, and model Russian helicopters and tanks—had arrived in our street. I never had a "best friend." My best friend was Grandmother, who liked to tell me, "You are a lion in the house but a mouse outside." Years afterward, I met a girl who had known me as a child in Kabul. "Whenever I came to see you, you just refused to talk to me," she told me. Perhaps the problem was that I was used to being alone or with adults, and I expected other children to behave like adults. I found it boring if they behaved like children.

I was curious about only one girl in my street. Her father, who had a long beard, didn't want her to play with any of us, and she would stand away from us outside her house, always wearing a heavy scarf that covered most of her head and face. I asked Grandmother about her. She told me that the father had four wives and rarely allowed them out of the house. "They're all stupid in that family," Grandmother told me. "Don't bother trying to talk to any of them." Later, when I could understand what the word meant, she told me the father was probably a fundamentalist.

The only girl I played with was Khadija, a girl who lived three doors away from me and whose mother was a school-teacher. Like me, she usually preferred the company of adults, and that bound us together. She had a lot of dolls, and she would bring them to my room where we would put them together with Mujda and organize a birthday tea for her. I had seen adults offer tea to their visitors many times. It was a tradition that a cup should be handed to a person who entered your home. I had seen adults insist again and again, while the visitor continued to refuse politely.

Once, Khadija told me we would use a *burqa* for a game. I didn't like the idea, but Khadija insisted, and she dragged me to the house of a neighbor where she said we would find one because a relative who was visiting that week always wore one outside. We would have great fun with it, she said.

Khadija, who had also recruited a couple of friends of hers, was in charge. "Zoya, you put on the *burqa*," she told me. "You will pretend to be a ghost, and you will run after us making a lot of ghost noises. We will run away from you, and you have to try and catch us."

It did not sound like fun to me. Although I was afraid of crossing the yard at night to go to the toilet, I did not believe in ghosts. But Khadija and the other children made me sit down and dropped the *burqa* over me. I had no idea how to put it on, but the girl who lived in the house knew because she had seen her relative change into it. It was big and heavy. I couldn't breathe or see properly, and I swayed from side to side as I tried to get up.

"Go on!" Khadija shouted. "Chase us!"

I tried to pick up as much of the thing as I could in my hands, and took a couple of steps. I couldn't even see where the other children were. I felt someone give me a big push in the back, and I fell flat on my face, the mesh over my eyes slipping to the side of my head.

I could see only darkness. I had become blind. "I can't see! I can't see!" I shouted over and over again until they helped me fight free of the *burqa*. "I hate this game," I said. "I don't want to wear it again."

KHADIJA TOLD ME a lot about the school where her mother taught, and although I wasn't keen on the idea of being with a lot of other children, I would have liked to find out what it was like. In the streets, I had often seen the girls in their uniforms, laughing and ringing their bells as they rode their bicycles to school. I would have liked to ride a bicycle of my own, carrying my books the way they did.

But my parents made me study at home. Every evening when Father returned home, he would impose the same ritual on me. He would call me over, sit me on his knees, hug me close, and kiss me. He was very affectionate with me, and I hugged him more often than Mother. His beard prickled; he was lazy about shaving and did it only once a month.

He always asked me what I had done that day and set me a task for the next. The task was always the same: I had to write a few lines in Persian, the language we spoke in my family, on a subject he had chosen, such as Spring, Kites, or Respect for My Elders. The next evening he would want to read what I had written, and correct my mistakes.

It was all he ever wanted to talk about with me, so I stopped wanting to see him. I came to dread his return and his request for me to bring him tea so that we could discuss the homework he had given me. If I hadn't done it, I would hide and hope that he would forget. Or I would rush to Mother and ask her to help me write something quickly.

But he could tell when I had written something in a hurry. "This is not very interesting," he would say. "I wanted to see you thinking hard, to see that you are reading things with attention, not just writing the first thing that goes through your head."

The same subjects kept coming back to haunt me, and there was a limit to what I could say about Sand, or The Ventilator. Every day I hoped that he would arrive later and later and that I would be asleep in bed by the time he came back. Years later I was to regret that I had ever wished such a thing.

The closest I came to seeing what school might be like was with Sima, a woman teacher that Grandmother, with the money her husband had left her, paid to come to the house. For a while she came three times a week. We would sit on the floor, Sima would take out her book and her knitting, and right through the

39

two or three hours we spent together she would never stop knitting, the needles darting and clicking in front of me.

With Sima I never got much beyond learning to read and write in Persian and doing arithmetic. She never seemed to care whether I learned anything or not. She would give me a Persian book and tell me to copy ten pages of it. It didn't matter to her that I didn't understand the meaning of what I was copying.

Whenever she saw me getting bored, she would start chatting with me and telling jokes. Although I was only a child, I knew this was not good for me. Soon her visits became very irregular, and then she stopped coming altogether. I did not miss her. My parents tried a few other teachers, but two years after I had started having lessons at the age of seven, nobody came to teach me anymore.

It was only much later that I found out why my parents refused to send me to school. They feared that the Mujahideen might set off bombs at the schools or at other buildings on my way there, and in any case they had no patience for the way the schools were run. The subjects were all taught according to the guidelines laid down by the puppet regime, and many of the books given to the children were translations of Russian textbooks. My parents thought the children learned more about Russia than about Afghanistan.

Grandmother insisted that I study at home as much as possible. She was so adamant about this that she quarreled often with a neighbor who sometimes called on us. "I'm not saying that she should grow up to become a housewife," the neighbor would say, "but this girl should know something about cooking and keeping a clean house. Otherwise how can she live with a husband?"

"Wrong," Grandmother would retort. "She should do the work that boys do. There's no future in cooking and cleaning. Education and knowledge, that's what she needs. Zoya, get out of the kitchen and go read a book."

I never did learn how to cook.

I HAD BEEN helping Mother with her work whenever she would allow me to for little more than a year when one afternoon, after we had finished our lunch of beans and rice and I had wiped clean and put away the *destarkhan*, the plastic tablecloth that we spread on the carpet, Father gathered us around him and said he wanted us all to listen to something.

He put a tape in the cassette player and said that a friend had recorded this when he had gone to the main prison in Kabul to see some official notices that were displayed there for the public. As far as I could tell, it was just a list of names—many, many names, so many I could never have counted them.

But the way Father, Mother, and Grandmother froze as they sat on the cushions made me realize that something terrible was happening. The list went on and on.

A shout from Father suddenly broke the monotony of the list. He shouted bad words about the Russians, about the puppet regime. *"Watan frosh,"* sellers of the country, he shouted, as weli as other words I had never heard him use. I started to ask what it was all about, but I was told to be quiet.

The tape ended, but the adults just sat in silence, until Grandmother got up and said she was going to bed. Father and Mother went on sitting there, not talking and taking no notice of me, so I left them to follow Grandmother.

I found her praying aloud, saying over and over again, "Allah, bless the martyrs who try to free our country." She explained that my parents knew several of the people on the list, that they were politicians, writers and poets, professors from the university where they had studied, courageous people who had taken a stand against the invaders.

"Listen, daughter. All these people, they wanted to get the Russians out of our country. And the Russians tortured them and killed them," she said. It was the first time Grandmother had told me the truth about life under the Russian occupation. That night I fell asleep with the sound of her praying in my ears.

What happiness there had been in my house vanished from it. The next day Father went out of the house early, and Mother was silent. I spent most of the day with Grandmother, whom I watched as she took her clothes out of the rusty iron trunks that she kept along the wall of her room, and then put them back again.

"Think of the mothers and the wives of these poor people," Grandmother said to me. "Now they have only the graves." I did not tell Grandmother, but I thought about how I would feel if the names of my parents appeared on such a list when I was sitting at home waiting for them to come back.

For the first time I realized that people were being killed for their ideas, and fear entered my house as never before. My parents worried that the Russians had recruited spies among the people of Kabul and that even our neighbors might betray them. Again and again Mother told me not to speak to anyone about the work we had done together.

From that moment on, Mother left the house even more often. Father was sadder than I had ever seen him. When I asked whether I could go outside to fetch something from the shops, I was told it was too risky. "Don't go out on your own. There is danger outside" were words that became stamped on my brain.

A FEW WEEKS after Father had brought the tape home, after I had badgered him for permission for days, I was allowed to go out for an afternoon to visit Aunt Naseema, who lived in the city. She came to fetch me at home, and later persuaded me to stay the night, although I wasn't able to warn my parents that I would be staying longer than expected. I knew that I was heading for trouble. From my earliest childhood, I can remember Father saying time and time again: "Where you spend your day is up to you, but at night you must always be in your own house." But Aunt Naseema said that she would have a word with Father when she brought me back home.

The next morning Aunt Naseema accompanied me as

promised. Father kissed me in front of her, but I could tell he was angry. He asked Aunt Naseema where I had been, and as soon as she had left, he called me into his room and asked me in a quiet voice why I had disobeyed him.

"Aunt Naseema said it would be all right, that she would explain," I stammered. I felt very small and stared at my feet, not daring to look him in the eye.

"Don't you realize you are living in an occupied country? Now listen. I have never slapped you before, but now you have done something that I forbade you to do. What should I do to you?" he asked.

I said nothing. I had never seen him so angry. I thought of running to Grandmother. Perhaps I could hide behind her and use her as a shield. Surely I would be beyond Father's reach.

"What should I do to you?" he repeated.

The slap burned a hole through my cheek. But I stood still as a second slap hit me on the other cheek.

When I went in tears to find Mother in the kitchen, she dropped what she was doing, hugged me close to her, and left me to find Father. I heard them arguing. Mother told Father that I was only a child, I had understood I had made a mistake, and there was no need to slap me.

But Grandmother said he had been completely right to punish me. She told me I could have been hit by a bomb in the street, and no one would have known.

PART TWO

The Bleeding Wound

Chapter Four

SOON AFTER my eighth birthday, I finally solved the mystery of what Mother was up to. Over the past few weeks she had looked increasingly exhausted and pale, until one evening, just after she had entered the house, she suddenly put out her hand to steady herself against the wall and then slowly dropped to the carpet in a dead faint.

I rushed to the kitchen to find her some lemon or orange juice. Then I boiled some water with the electric rod to make her some sweet black tea. Then I sat quietly with her while she recovered her strength, and I massaged her feet to help her relax.

When she went to bed later that evening, I went to her and lay down beside her. Mother kissed me, ran a hand through my hair, and then started rubbing oil into it. Her gold ring grazed lightly against my scalp. I turned toward her and asked the questions I had asked so many times before: "Mother, where were you today? Why are you never in the house with me?"

This time there was no throwaway answer. She stared solemnly at me with her big black eyes. She began by reminding me of one of our neighbors, a woman who often came to us in tears because her husband beat her. "Well, you have to know that here in Kabul there are women who are beaten or treated very badly, not just by their husbands but also by the soldiers. I have met these women, I have seen the pain in their lives, and I believe we should help them. And then there are many women who are desperate because their husbands have been killed or put in prison by the government, not because they did anything wrong but just because they want their country to be free."

"So how can you help them?" I asked.

Mother stopped massaging my scalp. I touched her hands so that she would continue. "It is very difficult," Mother said. "We are not magicians. There are several of us, women who are all part of a group that tries to help women and to bring peace to our country. That's what the papers that we write and distribute try to explain, that the people of our country have the right to decide their own future, and we have to fight, without violence, against the wild Russians who want us to be their slaves."

I was intrigued. In the films I had seen at the Barikote Cinema on the rare occasions when my parents took me there, policemen fought bad men with guns, and they often won. I thought it was easy to solve problems with a gun, and that to be a heroine in the real world, just like in a film, you needed to have one.

As Mother braided my hair into two pigtails, tying them with colored bands, I heard her pronounce for the first time the name of RAWA—the Revolutionary Association of the Women of Afghanistan. "It was created a few years before you were born, by a group of professors, students, and intellectuals. I was studying at the university, but I stopped because this work was much more important. You remember those papers you carried in your backpack? Well, those were written by RAWA. They were secret papers that denounced the Russian invaders, and said that people should resist them."

"But, Mother, what could happen to you? I am frightened for you," I said.

She stroked my hair. "No, don't be frightened. It is not dangerous. I am not alone, and there are other women doing more than me. And how can I sit in the house and do nothing? If we don't help the women who are suffering, no one else will."

After that she told me stories about people who had fought against fascism, racism, and other evils in many different countries. She told me about Malalai, an Afghan girl who had fought against the British when they sent their army to invade my country. During fierce fighting in a battlefield called Maiwand in July 1880, the soldier carrying the Afghan national flag was killed and fell to the ground. Malalai rushed to pick up the flag and declaimed a poem:

49

With a drop of my sweetheart's blood (shed in defense of the
  Motherland)
Will I put a beauty spot on my face
Such as would put to shame the rose in the garden.
If you come back alive from the Battle of Maiwand,
I swear, my sweetheart, that the rest of your days you will live in
  shame.

Still today, her words give courage to Afghan fighters.

Mother told me that Father was also involved in fighting the Russians and Moslem fundamentalism. He was not a member of RAWA—only women could be members—but he was part of another secret organization. I was never told which one, and he always refused to tell me what he was doing. In general he was not very talkative, and asking him how he had spent his day was certain to make him even more silent than usual. "I'm going to see a friend" was all he would say. Father never spoke to me about his ideas.

It was very late when Mother sent me off to bed. "Tomorrow morning, ask Grandmother to tell you about her marriage," she said to me before I fell asleep.

BUT THE NEXT MORNING Mother had other plans for me. I would have to wait until later to ask Grandmother about her marriage, because Mother wanted me to accompany her to the *hammam* (steam baths). I was so rarely allowed out of the house that I looked forward to the trip, even though I didn't like the way Mother would scratch my skin until it hurt. When I was smaller, we used to go to the *hammam* once or twice a month because we could not wash properly at home, where there was no heating, when the weather turned cold. But for some time now we had been going less often.

Mother packed my knapsack, and we set out. At the *hammam* there was a public bath where the women sat around talking, washing one another, or sitting perfectly still, but I didn't like it there because I was embarrassed to be seen naked by strangers.

Mother paid for a small separate room. She preferred it because it was even hotter than the common room. After we had undressed and stored our clothes in plastic bags so that they would remain dry, she sat me down on a little stool. She poured water over me and scrubbed me down with a rough cloth so hard that I shouted for her to stop. "My skin is so red I can't stand it!" I cried. But she showed me the black stuff that she had rubbed out of my skin and told me there was plenty more that she needed to get out.

There was a corridor outside, and while I sat in the room steaming, Mother would sometimes go out, wearing a towel, and talk to some women there.

I was so exhausted after the visit to the *hammam* that when we arrived home I paid no attention when I heard Grandmother telling Mother, "You must be careful. Things are getting more and more dangerous."

I went straight to sleep for three hours. I could not understand why women so enjoyed going to the *hammam*.

Much later, I found out that Mother used it as a place to meet her RAWA friends and to pass on documents to them—something she had hidden even from me. The *hammam* was one of the few public places where women could meet discreetly. No one would have thought of looking in a child's knapsack at the bath.

IT WAS ONLY that afternoon that I got to talk to Grandmother. As she made halvah for me with ground flour and sugar—the smell of sugar cooking filled the house, and I loved helping her to balance a big teapot on top of the mixture so that it would set properly—she told me the story of her arranged marriage and of the suffering it had brought her. She called her late husband "your grandfather," never "my husband." I had never met him. He died before I was born.

"One day my father—he was the mullah of a mosque in Kabul—came to me and said, 'You have to marry this man.' I was only thirteen years old, and I had never seen him before. My father never told me where he had found your grandfather. Perhaps he met him at the mosque. I think my father chose him because he had more money than most people. But he wasn't rich. I wasn't told anything about him, and the first time I saw him was on my wedding day."

Even at the engagement party, Grandmother was still not allowed to see him. The husband's family brought presents and cakes to her house. But the men and the women did not mix during the celebrations; they were kept apart in separate rooms. The rule was that the men ate first, and when they had finished, the women had to clear everything away and wash the dishes. Only then were they allowed to eat whatever food the men had left.

Two months later Grandmother saw him for the first time. She found out that he was twelve years older than she. On the wedding night he was kind to her. He was handsome and seemed to be a good man. Over the next four years, she gave birth three times.

In the end he turned out to be very traditional, very strict, and very selfish, and he treated her very badly. Grandmother said he insisted on good, expensive food. Every morning for his breakfast he wanted a fried egg with a glass of milk. But he never stopped to think that he was the only one having this kind of breakfast, and that Grandmother would have only a little bread and tea. He didn't think of the children either, didn't care that they needed proteins and vitamins to grow up properly.

When Grandmother told me that he would beat her, I asked her why. I thought she must have done something wrong and been punished, just the way a child would be beaten for making a mistake.

Grandmother shook her head. "It was different. Your grandfather was heartless. He would arrive at the house with a dozen friends, and he would order me to prepare lunch immediately. And it must be a big meal, with fine dishes. Once I said to him, 'I am too tired. I am your servant, but even servants are human. I haven't got the strength to do what you ask. And we don't have enough plates for everyone.' I said it in front of his friends, and he went to fetch his boots, big boots that he used for walking through the snow, and he beat me with them. That was the first time he beat me in front of his friends."

"Didn't they do something?" I asked.

"No, nothing. For them it was normal. They too beat their wives, and no one had the right to defend them. They beat their wives even when they were pregnant. When he'd finished hitting me with his boots, he ordered me to go and borrow some plates from the neighbors and to make the lunch. And in the afternoon, when his friends had gone, he beat me again. He kept shouting, 'Why do you come to me in front of my friends and ask for plates?' That night, when I got undressed, all my body was covered in bruises."

"Didn't you ask your father, the mullah, to help you?"

"I did go to see him, several times, but every time he just told me to put up with it."

"It's your fault if you stayed with him. You could have run away, but you never did."

"It's not as easy as you think. If I had run away, I would have had a bad name. A husband is the protector of his wife, and if I had divorced him I would have been left defenseless. And despite the beatings, I still loved and respected him."

"But you must be pleased that he is dead," I insisted.

"Don't say these things, daughter. I am very unhappy that he is dead. But he was called to Allah, and I'm sure he went to a good place."

Although she talked often about Allah, she never tried to make me as devout as she was. All she told me was that Allah could help solve any problems in my life. She fasted for Ramadan but told me this was not for children. She prayed five times a day, but she said it was not only praying that made a good Moslem. All I needed to do was to be good and kind to other people, and that, she said, was difficult enough. "And helping the poor is worth more than any number of prayers. In any case, Allah will know whether you are a good Moslem," she told me.

The only thing she demanded of me was that I be careful, when she was praying, not to pass by the prayer mat if there wasn't a cushion, or something else, placed between us. That was a sin.

Whenever we spoke about her husband, she would urge me to learn from her experience. Sometimes I heard old women talking about husbands beating their wives, and the old women would say things like "It is their right, and we have to accept that," or "This is our destiny, and we must bow our heads." Grandmother had no patience with these sayings. "For the women of my generation, things were different. But you must never tolerate what I went through. It was not human. I admit it, I made a mistake in putting up with it. You must have a good education. You must never be shy of expressing your opinions in front of men. Don't let them have power over you. And the man you choose as your husband must be well educated, he mustn't be ignorant, and he must promise to respect you as a woman."

FROM THAT TIME ON, I had only one ambition—to understand the books that Mother read and to study so that I could follow in her footsteps. Before, I had been bored by my parents' conversations about politics. Now I wanted everything to be explained to me. My parents told me that the KGB and the KHAD, the Afghan secret service, were arresting and torturing thousands of people.

My parents ridiculed the regime that had been installed by the Russians. "When it rains in Russia," a popular saying went, "the regime in Kabul has to open its umbrella." They also kept telling me that even I could do something for Afghanistan later on. We did not think much of a distant cousin of ours who had managed to emigrate to Canada to study there.

In 1986 Mikhail Gorbachev, the new Soviet leader, described Afghanistan as "a bleeding wound" and decided that his troops should begin planning to withdraw. He could no longer afford either the economic or the human cost of the war, which his country had fought to keep its influence over a strategically important neighbor. My parents were skeptical. They didn't think the pullout would happen anytime soon, and told me that the main problem after their retreat would be how to ensure a good future for Afghanistan. The Russians were counting on their men staying in power once they had left. Gorbachev installed Mohammed Najibullah as president three months after his "bleeding wound" speech.

I was prouder than ever to carry the papers in my knapsack, and I held Mother's hand more tightly. Gradually I started doing the same thing for Mother's friends, because a child would go unnoticed by any Russian patrols. Even if the soldiers had searched my bag and found the papers, nothing bad would have happened to me. I did not think of what might have happened to Mother.

I learned to lie when the other children in the street asked me where Mother was. It was a small lie: I told them that she had work to do and that no, I didn't know what this work was. I could be honest only with the children of Mother's friends inside RAWA, who sometimes came to visit us—children who like me would complain that their mothers were never home.

I no longer asked Mother why she did not spend more time with me. I knew that her absences did not mean she did not love me, and I was never jealous of her work. I came to realize that if she had to choose between me and her work, she would choose her work. I had no doubts about this, and I was proud that Mother would make such a choice.

When I look back on those days when Mother told me more about her work, I see them as the end of my childhood. I feel no sadness about this. I hated being a child. It was pointless. I wanted to grow up fast so that I could achieve something useful. I think that Mother wanted me to have an aim in life, and she may also have thought that, because her life was at risk, I should understand what she was doing.

Mother never did admit to me that there was danger in her work.

Chapter Five

IN FEBRUARY 1989, when I was eleven years old, the Russians finally abandoned Afghanistan. It was the end of nine years of occupation, and came three years after Gorbachev had spoken of the "bleeding wound." One of the biggest empires in the world had been humiliated, despite deploying a total of more than six hundred thousand troops over the years. Afghanistan had lost more than a million lives, and more than six million Afghans had been forced to flee as refugees—the world's biggest refugee population. We knew the wound would continue to bleed, because the Russians left behind the mines they had hidden in our soil.

This was no time to rejoice, and we had no peace after the Russians left. I knew there was no chance I would soon be going to school. The Afghan fundamentalist warlords fought against the Najibullah regime, raining death down on the city of Jalalabad. Months later they turned against Kabul, and soon they were hurling rockets at the capital even more intensely than they had done when it was in the hands of the Russians. At the rate they were bombing Kabul, Grandmother said, there would be little left of it for the victors to enjoy.

In the first months of 1991, the bombings were heavier than ever. One morning I saw Grandmother lifting boxes of clothes, broken kites, and a hundred other things out of the cellar. I thought nothing of it, but the next time the bombing started, I was made to go down to the cellar, and it was hours before we could come out again. I must have spent dozens and dozens of hours down in those two dark, damp rooms.

The entrance to the cellar was in the garden. It was nothing more than a hole in the ground, with some steps leading down to the cellar immediately below. As Grandmother pointed out, a bomb could have dropped right down the hole at any time, and the wooden board that we placed across it every time we went down there would have made not one bit of difference.

Grandmother had moved out of the cellar all the useless things we hadn't touched in years, and we tried to make the shelter, as we came to call it, as comfortable as possible with carpets, cushions, and *toshak* mattresses for us to sleep on. But it was still wet, very cold with no stove to warm us, and very dark. We painted the walls of the shelter, but they soon became spotty with marks of damp, and even the mattresses started smelling wet.

It was disgusting. The damp gave me a bad cough. I couldn't understand why I was not allowed to go and sleep in the comfortable bed that was waiting for me upstairs. Grandmother complained too. She was uncomfortable going up and down the steps, she protested, and her asthma and her rheumatism were getting worse and worse. The longer she was forced to stay in the shelter, the more she grumbled, and the more she argued with Mother.

In summer, the heat inside the shelter was stifling. Before, when it got too hot in the house, we used to put our beds in the garden and sleep under a big mosquito net. But now we were forced to stay underground, the only air coming from the cracks in the board that Father had put over the opening.

I hated the nine steps down to the shelter. I couldn't go out, I couldn't see the sun, I couldn't tell whether it was night or day. We had a couple of lightbulbs, but the electricity was often cut off and we would have to sit around a candle. Often we were stranded there for several days at a stretch. The only time I saw the sky was when I scrambled out to go to the toilet behind the house, but that was only for a moment because I always had to run straight back.

When there was enough light, I tried to read, or drew guns and soldiers and tanks. When it was too dark to see, I asked Grandmother to tell me stories. I helped prepare the little food that we could find for ourselves, often peeling potatoes by candlelight. In the evenings, if the electricity was working, we listened to the BBC Persian Service, or to Afghan songs, on the radio we had brought from upstairs. We didn't have a television.

One day we heard from some neighbors that an entire family, both parents and their six children, had been killed by a bomb landing right inside their shelter. The neighbors said it happened far away from our street. After that, I couldn't understand why we were spending so much time underground. I said to Grandmother: "There's no point. Whether we're in the shelter or in the house, if it's been decided that we must die, then we will die. So let's go upstairs."

I thought Grandmother would agree with me, she had been complaining so much. But she simply said there was less risk of getting hurt if we stayed in the shelter. So we stayed.

After hearing that a whole family had been wiped out so suddenly, I found it harder and harder to get to sleep. I was exhausted, but I wanted to stay awake. I clung to the stupid idea that if I were awake when the bomb came, I would be able to run away. I was convinced that the more I thought about running away, the less likely it was that the bomb would come. And if I closed my eyes and went to sleep, when I opened them again there would be no house, no parents, no Grandmother, nothing. Everything would have disappeared.

April 28, 1992. The black day that I can never forget. I was having breakfast with Grandmother when the radio announced that the fundamentalist Mujahideen, for once united, had taken power in Kabul. Far from rejoicing that the Russians had been defeated, Grandmother told me that a new, worse Devil had come to my country. There was a popular saying around this time: "Rid us of these seven donkeys and give us back our cow." The donkeys were the seven factions of the Mujahideen, and the cow was the puppet regime.

The guns on the mountains had gone quiet, but again I was imprisoned. Before, I had been confined to the shelter. Now, I was confined to the house, and I was allowed out only once every two weeks to go to the shops or to visit my friend Khadija, never for very long and never on my own. I was told to stay as far as I could from the checkpoint on the main road at the end of our street. The soldiers had wild-looking eyes and long beards. What I did see of my divided city filled me with sadness. Rival ethnic groups controlled the various neighborhoods. Only the dust in the streets was still the same, but there was even more of it. There were fewer children in the streets, and they didn't shout and laugh as they used to.

The shop where the Russian woman-soldier had offered me a chocolate had been destroyed. There was only one blackened wall left standing. So had the Barikote Cinema I used to go to as a child. The bombs had leveled entire neighborhoods. The museum where my parents used to take me had been looted; all the most expensive statues had been stolen.

I saw the Russian-built university where my parents had studied. It was gutted, all the windows shattered, the walls riddled with holes. The Mujahideen had even burned all the books from the rich library and looted its modern laboratories. They said the books had been supplied by the Russians, so they represented Communism and had to be burned. Alcohol was just as evil an offender, and the Mujahideen drove a tank over a thousand bottles to try to impress and frighten the people.

I saw more *burqas* than before. They looked like dead bodies drifting down the streets. Many women covered themselves with big scarves. The beautiful young women of Kabul no longer wore makeup or skirts; they tried to look old and wore only sad colors. Even I had to dress differently—Grandmother made me longer skirts than I had worn until then. She told me that the fundamentalists who had taken power beat women if they did not cover themselves enough.

Women started disappearing from the programs on the radio and on television. I thought how dangerous it must now be for RAWA to continue its work.

I used to get several presents for my birthday, but the only birthday present I got during this time was a very small red knife, which you could open and close. Grandmother gave it to me wrapped in some paper, kissed me, and said, "I'm sorry, daughter, I don't have anything else to give you." The knife was hers. She had had it for as long as I could remember. I asked her when I would be able to go to school, but she had no answer for me. By this time, she was too weak and sick to go out.

I no longer saw the girls in their uniforms on their way to school. Somebody told Grandmother about a mathematics textbook for boys that had shocking exercises in it. One exercise was

"If there are ten Jews and you kill five of them, how many are left?" Another was "If you have ten bullets and someone gives you five bullets, how many bullets have you got?"

In the mornings, in the streets near my house, I saw groups of boys going to the *madrassa* religious school, where Grandmother told me they would sit cross-legged over the Koran and read from it out loud for hours on end, bending backward and forward like trees in the wind. The boys had to learn much of the Koran by heart, and studied the life and beliefs of the Prophet Mohammed as well as the *Shari'a*, or Islamic law.

A mother Grandmother knew had sent her son to one of these schools and was now complaining to her that he had changed into a stranger almost overnight and was refusing to obey her or show her any love or affection.

Grandmother said that some mullahs at these schools committed bad, indescribable sexual crimes against the children. It was a sin, she said, and it was being committed in the name of the Koran. I could not guess that in a few years' time, the name Taliban, which meant simply a student or seeker in a *madrassa*, would become famous around the world and send us back to the Stone Age I had read about in my books.

Over the following months, I spent days on end stranded at home. My parents were as busy as usual, leaving me alone in the house with Grandmother. I would try to read, but I preferred to learn from Grandmother of the brutalities committed by the fundamentalist Mujahideen. People were shot in the street for no reason, or simply vanished. I learned of a checkpoint where a commander of the Hazara tribe had built a pile of human eyes in the street. The eyes had been gouged out of men of the Pashtun tribe, my tribe. The Pashtuns started doing the same thing, and it turned into a savage competition. I remembered the true story Grandmother had told me years before of the Pashtun king who had built a tower with the heads of Hazara men.

I heard that fighters of the Hizb-e-Wahdat faction had rounded up many women and locked them into the Barikote Cinema. They forced the women to strip naked, and gang-raped them every night.

I heard of a commander who would beat and torture the people who had to pass through his checkpoint on the main road between Jalalabad and Kabul, and rape the women. Once the commander stopped an old man and asked him, "How much money do you have in your pocket?" The old man replied that he had a thousand rupees.

"Give me this money," the commander said. "But I have no other money," the old man pleaded. "I have to go to my family in Kabul. Please leave me something."

The commander shouted, "You refuse me? You should know me better!" And he pushed the old man to the ground, drew his whip, and lashed him time and time again, laughing as he did so.

The commander was famous across Afghanistan because he had a "human dog." The human dog was a wild man, very dirty, with long hair and a long beard, full of lice. He was said not to have washed for months, and the commander kept him on a chain. When the commander ordered, "Dog, come here and bite!" this man would come crawling up and sink his teeth into the victim.

When I saw these illiterate criminals on the televisions in the shops, or their pictures in the newspapers, I imagined blood on their hands and on their faces.

I heard that the victorious Mujahideen warlords had invented a new form of killing, which they called "the Dance of the Dead Body." The soldiers would brandish a sharp knife, making a great show, and cut the victim's head off. Then they poured boiling oil on the neck to stop the blood flowing and let the body drop to the ground, where it would shake all over. The Mujahideen found this very entertaining. They were so happy with this show that they would dance until the victim finally lay still.

One day in 1992, some soldiers stopped a pregnant woman who was on her way to the hospital to have her baby. At gunpoint they forced her out of the taxi. They told her that they had never seen a fetus inside a womb before and wanted to know what it looked like. Then they raped her. Her body, with the stomach split open, was not found until a few days after she died. The Mujahideen also liked to blind people by ramming a lighted cigarette into their eyes or kill them by hammering a six-inch nail into their heads.

These men just wanted to spread fear among our people, and there was nowhere to turn for protection—no law, no justice, nothing. The violence was institutionalized by the new government. Saying that it would impose Islamic law, it decreed punishments that not even the Russians had meted out: amputation of feet or hands, whipping and stoning to death. The political prisoners who had been jailed by the former regime were freed from their cells, only to be replaced by anyone who dared to criticize the new authorities. The new prisoners were tortured, often to death. The government instituted public executions for convicted murderers.

Then the bombing started again. In August 1992 nearly two thousand people were killed in Kabul when a Mujahideen faction rained shells down on the city.

I told Grandmother the Mujahideen must have come from the Kabul Zoo. "No," she replied. "Don't compare them to animals. Animals are innocent. They never do the things that these men do."

Chapter Six

THE LAST TIME I saw Father was on a sunny morning when he bent down to stroke my hair and kiss me on the cheek, and the beard he had trimmed a few days earlier prickled my face. He put on his coat and then slipped on the shoes that were outside in the yard, just by the door to the street.

At first I was told that he had gone away for work and that he might not be back for a few days. Then, when he had still not come home, that he might be away for some weeks. But after a while I guessed, from the tears I saw in the eyes of Mother and Grandmother, that all this was a lie. I thought of the many times I had hoped Father would return late to the house because I hadn't done the homework he had given me. Now the wait was too long, and I felt guilty for having wished his absence.

Mother would try to hide away from me to cry in secret, but I could always tell that she had been crying. I stopped sleeping with Grandmother and shared Mother's bed, and late at night I could feel her body shaking slightly as she wept when she thought I was asleep. I pretended to sleep. I did not want her to know that I was listening to her. I could no longer smell perfume on her. She had stopped using it.

For days we did not sit down together for meals. Sometimes Mother and Grandmother would forget to feed me, and I would go to the kitchen to help myself. They were so kind to each other, united by a common knowledge that they did not want to share with me. We had no visitors. Perhaps because they wanted to protect me, no relatives called at the house. I wanted to ask Mother and Grandmother what had happened to Father. Was he in prison? Had he fled Afghanistan? Was he lying wounded in a hospital? Was he dead? I thought I understood what had happened, but I was not certain. I did understand, however, that I should not discuss it. I respected their decision not to tell me anything.

No one went into Father's study, and the books remained on the shelves, their pages unopened. I knew I would never again feel Father's hand on mine as he helped me guide a kite.

We lived off the money Grandmother had inherited when her husband died. Whenever Mother was at home, I stayed by her side and did not go into Grandmother's room as often as I used to, because I wanted to show her that she was not alone. Mother did not have the patience to play childish games with me. She had to continue her work. But she started taking me with her. I accompanied her to some committee meetings, but one day I made the mistake of repeating something I had heard to a friend, and she left me at home after that.

I WASN'T EVEN in Kabul when Mother disappeared some time after Father. I had gone to visit a friend, Shaima, in a small town near Jalalabad. Shaima and her family did their best to make me happy, but I could not stop thinking about Mother. I lay awake at night worrying about her. I pictured her sick in bed needing help. And I remembered that Father never liked me to stay away from the house for very long. Shaima and her parents asked me to stay on, but after four days I left, anxious to go back to Kabul and Mother.

But when I got back home, the house felt different to me.

I found Grandmother in bed. She looked sick. She had tied a scarf around her head the way she did when she had a bad headache. I looked into her eyes and saw that they were bloodshot. She had been crying.

She didn't give me time to ask her anything. She gestured for me to come closer, reached out toward me, and put her hands on my cheeks. Her hands were burning hot. She had a temperature. She pulled me down so that she could kiss my eyes, the top of my head, my hands. Then she drew me to her chest in an embrace.

71

My head buried in her breasts, I heard her say as she began to weep again, "Daughter, what will become of you now?"

I pulled back from her in alarm and shouted at her, "What has happened? Tell me what has happened!"

But her only answer was to cry louder. Then she said, "Your mother, your mother." I fled to my room to be alone.

Later that day, when I asked her where Mother was, she did not answer. Over the days and weeks that followed, I asked her perhaps only two or three more times, never expecting her to answer.

I locked myself in my room. I refused to see the relatives who called at the house. I didn't want to hear their conventional expressions of sorrow or pity.

I huddled up on my bed to make myself as small as possible, or marched up and down like a robot from one end of my room to the other. I thought of the perfumes still on Mother's table, that I would never smell them on her again. I would never feel her hand, with the gold engagement ring, massaging oil into my hair.

I felt that I had lost everything. I could still picture before me the smiles on my parents' faces, and the way they would look at me with tenderness and love. I wished I could have spent longer, much longer, looking into their eyes the last time I saw them. From those days on, and still today, memories of Mother, of Father—a game of blindman's buff we played together, Father checking on my homework—will come back to me like a film and make me start.

I began to write letters to friends, but I gave up after a few lines and tore up the paper. I put on a cassette to listen to some music, always switching off after only a few seconds.

I felt I had loved my parents more than my own life, and I thought of committing suicide. I had heard of many girls killing themselves when they had lost their families or after they had been raped. I thought it was the easiest way available for me to get rid of everything—the war, the Mujahideen, the killings, everything. But after a few hours I felt ashamed of thinking such thoughts. I was too young to give up on the world. My parents would have disapproved. Suicide was a sign of weakness. It ran against all they had taught me. I would have been throwing it all away.

The only people I agreed to see were some friends of Mother's from RAWA, who came to my door, said something brief, and left me in peace. They did not say empty words like "We are sorry." They told me that although I had lost my mother, they would try to help me the way a mother would. They told me that I must realize I was lucky to still have Grandmother and other people like my parents' relatives, because so many girls had lost everything, and there was no one to help them bury their mothers or their fathers. They urged me to think about the others who were suffering and to turn my grief into strength. I respected them for that. They took care of Grandmother and brought food for our meals.

It was a long while before I found out that my parents had been killed on the orders of the fundamentalist Mujahideen warlords, like thousands of other people. I cannot say what I know about their deaths, or when they took place, because this would be too risky for me. We were never given their bodies, and no funeral was ever held for their passing. Grandmother said the warlords had robbed us not only of two lives but also of two graves at which we could mourn. Still today, there is no grave at which I can mourn my parents.

One night soon after their disappearance, I swore that I would avenge them, not only my parents but all those people who had been killed without anyone knowing where, how, or why they had died. I would not avenge them with a Kalashnikov but by fighting for the same cause Mother had fought for.

"HAVE YOU HEARD? I can't think of anything worse!" Khadija exclaimed as she rushed into my house. She was wearing a scarf, and there was fear in her eyes.

It was the summer of 1992, a couple of months after the Mujahideen had taken over Kabul. Without stopping to take off her scarf, she took me into my bedroom and closed the door behind us. Her story came out in fits and starts as she marched up and down my room, unable to stand still. A Mujahideen commander, escorted by armed men, had burst into the home of a beautiful eighteen-year-old girl called Naheed in the middle of the previous night. She was the daughter of a shopkeeper and lived in a block of flats in Mikrorayon, an eastern neighborhood of Kabul that was wealthier than ours. Perhaps the soldiers had heard about her from one of the old beggar women they paid to work for them as spies.

They demanded that Naheed's father give her to them so that she could marry one of the soldiers. The father refused. "Leave this house," he told them. "How dare you come here in this way in the middle of the night? Let the soldier send his parents to me, and if my daughter agrees, I will give her to you."

But the soldiers refused to do as he wished, and they tried to catch hold of Naheed. She somehow found a moment to run out onto the balcony of the fifth-floor apartment and throw herself into the air.

Khadija paused to stare at me. "You understand? They wanted to take Naheed by force, and she had to kill herself." I noticed that she did not dare to use the word *rape*.

"You know what it means?" she went on. "We'll have to wear the *burqa*. We can't go out anymore. It's too dangerous. And even in our houses we are not safe. At any time a Mujahideen could beat the door down with a Kalashnikov and take us away."

I had heard and lived through so much suffering that all I could say was "Why are you telling me all this? What can I do?"

"Who else am I going to tell? You think I can shout this in the street?" she shot back. Then she hugged me tightly and left, saying only, "I must go to my house."

I found Grandmother cooking rice in the kitchen and told

her all that I had heard. Her eyes filled with tears, and she started praying aloud. I had never seen her pray so fiercely. When she had finished, she told me, "It means that we are close to the Day of Judgment."

"What is the Day of Judgment?" I asked, stumbling over words I had never heard before.

"It is a day on which everyone has to go before Allah, and you have to tell Him all the good things and the bad things you did in your life. And then He decides whether you go to Heaven or to Hell. In Heaven, there are rivers of milk and lots of fruit trees, and you can eat anything you like. You don't even have to ask for anything. You just think of something you want and it comes immediately."

"And what happens in Hell?"

"There's a huge fire, and there are two kinds of pain: either you burn and die immediately, or you burn for a long, long time. Or the bad people are forced to sit on giant thorns. The bad people are given only very little toast to eat, and no water."

Naheed had committed suicide, which was usually considered a sin, but because she had been under threat, she would surely be forgiven, Grandmother told me.

She told me that to go to Heaven, I had to be honest and kind, help the poor, and respect my elders. But I never did understand why Grandmother, when she heard of a new tragedy, always said that we were being punished for our sins. I used to feel I had done something wrong, that all my family and my neighbors were guilty of something, but I did not know what. I asked them all, and my friend Khadija. But no one knew the answer.

THE DAY AFTER Naheed killed herself, some RAWA friends of Mother took me to see her. Her father had put his daughter's body on her bed soon after her death. He had wanted his friends to help him carry it through the streets to show everyone what the Mujahideen had done to her, but some soldiers stopped him. When we arrived, many people were crowding around her body, but they let me get close to her. I saw that she was dressed in a white sheet. She had an oval face and high cheekbones. Her skin was almost yellow. There was no trace of blood. Someone had tied a piece of string around her head and under her jaw to keep her small mouth closed.

I didn't touch or kiss her, but I spoke to her under my breath. I promised her that I would bring the people responsible for her death to justice and that I would make sure they were punished for the terrible thing they had done. It was no empty promise— I knew that one day I would fulfill my promise to her.

Not even the Russians had done this to us. We heard that one man who had seen Mujahideen soldiers arrive at his home to steal his daughter had killed her before they could take her away. I imagined my fate if the Mujahideen, with their horrible faces, came for me during the night.

The violence that the Mujahideen inflicted on the women of Kabul and all the reports of tortures and killings affected

Grandmother a great deal. She was devastated by the knowledge that a nail could be driven into the skull of a human being in the name of her Moslem faith, and that a girl like Naheed could die in such a way.

She was not the same Grandmother I had known throughout my childhood. I often heard her crying, and when I wiped her eyes with tissues, it seemed to me that every time there were fresh little wrinkles around them.

She lost weight, and when I saw her eating like a bird, drinking only a glass of milk in the evenings, I told her that she must eat more or the medicine she was taking would hurt her. "I don't want anything now. Later. I will eat later," she replied.

I tried to comfort her. "Don't cry. Why are you crying? You are only making me more sad," I said.

After some time, she stopped crying and said to me, "If you are alive, then I am happy. That is all that matters."

She stood up, wincing at the pain in her bones, and made her way to the kitchen, where she busied herself for a while. When she came back to me, she was holding a frying pan in her hand, and smoke was coming from some little brown seeds that she had apparently been cooking. She moved around me, the wisps of sweet-smelling smoke touching me as she said a prayer to ward off evil. "Bless her, take care of her, save her life," I heard her say.

The Koran, which until then used to lie on her table most of the day, was always open now on the carpet, and Grandmother prayed from morning until night. Her beliefs changed. All she said to me was that yes, she did believe in Allah, although I should also count on my own strength to achieve things.

But at night I heard her pray to Him in a voice full of self-pity and bitterness: "Oh, protector of the entire world, I am a Moslem, Afghanistan is Moslem, and now these criminals have come to kill us. Allah, what sin have we committed that you are allowing this? Allah, please help us."

Until then I had accepted Grandmother's verdict when a relative or a neighbor died: "It is in the hands of Allah. For some reason that we do not know, He called this person to Him. It is all for the best." Now, when even Grandmother was questioning her God, I no longer knew what to believe.

PART THREE

A New Name *for* Freedom

Chapter Seven

I WAS GIVEN little notice to prepare for exile. I was fourteen years old in 1992, and unknown to me, Grandmother had for weeks been talking to RAWA members who had come to see her at our house about the dangers facing us in Kabul and about the need for me to get a proper education. Both Father's and Mother's relatives had offered to take care of me, but I wanted to stay with Grandmother and would do whatever she decided.

Grandmother wanted us to leave Afghanistan because, especially since the death of Naheed, she feared that at any time a Mujahideen would burst into the house and abduct me. She had also come to realize that she was too sick to take proper care of me.

Grandmother didn't consult me about her plans. She just sat me down in front of her on the floor and told me what I must do. We would leave the following morning, she told me, and we would travel across the border to a town in Pakistan so that I could start school.

She was sorry to leave Kabul. Her life was there, as were the children, although she saw them only from time to time. "If it was for me, despite everything that has happened, I still love it here and I would stay. But we have to think of your safety and of your future," she said. RAWA was offering her a chance to leave, and she realized that she must take it.

I didn't argue. I knew instinctively that whatever she decided for me was right. I simply asked her, "What kind of a school will I go to?"

"I don't know, but it will be good for you. We will start a new life in Pakistan," she replied. "And if anybody asks you tomorrow what you are doing, just say that you are traveling to see your relatives."

Grandmother, usually so slow when she moved about the house, busied herself at what was for her a breakneck pace. She rolled up the carpets and wrapped things up in paper so that they would not get dirty.

I wanted to take many things with me—my books, my toys, everything I loved in my room. But Grandmother told me we were in a hurry and could take very little with us. She packed two bags for me, and they were filled mostly with clothes, both for summer and winter, because we did not know how long we would be gone. Apart from the clothes, I was allowed to take only my doll Mujda and a few poetry books. I had to leave behind a big furry bear that Father had given me years before.

That night I tossed and turned as I tried to find sleep in Grandmother's room, wondering when I would next spend a night there. I was going to a country I didn't know, where I knew no one and didn't even speak the language. I too was sorry to leave, but if Grandmother, the only person I had left, was to stay with me, then I could face up to it. And if going to school was the only way for me to continue Mother's work, then I was ready to leave. Again and again I had told Grandmother that I wanted to start work immediately, but she always answered, "First, finish your education. Then you can think about work." The doubts and the questions went round and round in my head, and I got only three hours' sleep.

We were up at dawn with the call to prayer from the mosque. After her prayers, Grandmother carefully wrapped her Koran in a cloth and put it in her bag. She packed some biscuits and water for the journey.

My friend Khadija was the only neighbor I said good-bye to. I had no present to give her, and there was no time to get her anything. I told her where we were going. "You and your family should leave too," I said. "It isn't safe for anyone here."

"My parents keep saying they want to go very soon," she replied, "but I have no idea where we will end up."

At nine o'clock a woman from RAWA knocked at the door. She was wrapped in a big scarf, and her face was completely hidden save for her eyes.

Seeing her standing there, I remembered the fright Mother had given me when she came to that door in her dirty yellow *burqa*.

"A car is waiting for you at the bottom of the street," she told Grandmother. "The driver is a good man who has done a lot of work for us. You can trust him."

While they waited outside, I followed Grandmother as she went from room to room, checking the windows, locking the doors—my room, Father's study, my parents' bedroom. Grandmother was close to tears, but I did not know how to comfort her. All I could think was that I must take a good look at every room. Neither of us spoke.

A heavy bag in each hand, I walked next to Grandmother to the end of my street and got into the car. I said good-bye to the mountains around my city as well—they were my friends. It won't be long, I told myself. I'll be back in a few months' time.

I STARED AT THE BIG SIGN. "Country School for Girls," it read. The school was a big one-story building painted white, blue, and green. I was a teenager, and this was my first-ever day at school.

Grandmother and I were exhausted. We had been traveling for two days and two nights. We were driven across the Afghan border, our hearts in our mouths every time we had to cross a checkpoint on the road, to Peshawar in Pakistan, and from there we took a train to the town of Quetta, where we knew another driver would meet us at the railway station. It was only after we crossed the frontier that Grandmother told me I would have to live at the school in Quetta day and night and that she would go to live elsewhere in the town. I protested and cried, but in vain. If she had told me before we left the house that we would be separated, I told her, I would never have agreed to leave Kabul.

The gates of the school swung open for us, and we drove into the courtyard. A group of schoolteachers came to greet us, and girls of different ages in blue-and-white uniforms swarmed around us, staring at me.

I was near to tears but I forced myself not to cry because I would have been ashamed to be seen by the other children. "Where is Grandmother going?" I asked. "I am going with her. Grandmother, I don't want to leave you."

One of the teachers, a tall, graceful woman with a long neck, went up to Grandmother and kissed her hand. She was dressed in the *shalwar kameez*, like many Pakistani women—trousers and a long shirt that reached down to her knees. It was only later that I found out that the teachers, who were all Afghans, wore these clothes rather than more modern clothing so that they would go unnoticed in the streets.

Grandmother touched the teacher's cheeks and kissed the top of her head. It was a mark of respect typical of old women.

The teacher told me her name was Hameda. She smiled at me. "Don't worry, you will like it here. You will stay here, and we promise you that you will see your Grandmother soon."

"You will come to see me on Friday, the day of rest," Grandmother promised.

Hameda was gentle; she did not force me to let go of Grandmother. I hadn't expected a teacher in a school to be so kind to me. I had always heard that they were very strict. Hameda introduced me to some of the children, although I asked her not to. She told them that they should help me with my studies. I was too sad and confused to speak to them, or to remember their names.

I didn't go to class with the other girls that morning. I insisted that Hameda allow Grandmother and me to sit in a classroom all by ourselves at first while I looked at some books Hameda had given me. But after a couple of hours, I went up to her and said, "All right, Grandmother, you can leave me here."

The way we had our lunch of boiled beans, fried eggs, and rice was strange to me because I was used to taking my meals alone with Grandmother, but here all the girls sat together on the carpet. The teachers sat with us, eating the same food we did. We were all treated the same way.

When some girls came up to me after lunch and asked me to play with them, I said that today I didn't want to. I found a place in the yard where I could sit on a big stone and be alone, as far away from the other girls as possible.

That afternoon Hameda, who was the teacher of the Pashto language, which is widely spoken in Afghanistan, called me to her room. "I hope you will be happy here," she said. "I know that this is your first experience of school, so it may take time for you to get used to it. But I just want to tell you a few things about how this school is run."

She told me that the girls at the school were from all over Afghanistan. There were girls from the Pashtun tribe, which was my tribe, as well as the Hazara tribe and others whose names I had never heard before.

"But you must never, never, ask the other girls which tribe they come from," Hameda said. "Don't laugh at other girls because they are different from you. Some girls can't speak Persian, some speak it with an accent that will seem strange to you, some look very different from you. You should respect these differences. Where they come from does not make them better or worse than you. Never use violence against them, don't get into fights and pull their hair. Treat them all like your friends, like sisters."

My grandmother had entrusted me to the school, she told me, and RAWA would pay for my education there. But by the time I left the school, I must be a different person or the time spent there would be wasted. Class was more than sitting on a chair in front of the teacher for a few hours. It was up to me to do my homework and get good marks, up to me to study not because the teacher asked me to but for my own good. I should see the teachers not only as friends but also as mothers.

"If you are ever homesick," she continued, "try not to say in front of the other girls that you want to go home. All the girls at the school have families that are in Afghanistan, and some of these girls haven't seen their families for months, in some cases for years. You will only make them feel worse. If you don't respect any of these rules, I will have to be strict with you. We don't like to beat the girls here, but if they have done something serious, we have no alternative.

"Do you have any questions?" Hameda asked when she had finished.

I said that no, I had nothing to ask. I felt too shy and too confused to talk to her.

Before going to bed that evening, I was told to wash my hands and my feet and to brush my hair. Then I was shown to the dormitory. It was a long, narrow room, with bunk beds for the school's sixty girls. My bed was on top, in a corner. I liked that. I would be alone up there.

The lights went out at eleven o'clock, but I could not sleep. I thought of what Hameda had told me, of Grandmother and of our house in Kabul. I had accepted that I must leave my home, but now I was forced to abandon Grandmother as well. I was quite alone in bed without even Mujda to cuddle, having left her in Grandmother's care because I was afraid that the children at the school would want to touch her and that she might get hurt.

I buried my head under the blanket and tried to cry without making any noise, hoping that none of the girls would hear me. But Sajeda, who slept in the bed just under mine, did hear me, and she climbed up the ladder to sit beside me. Her legs hanging in the air, she pulled at the blanket covering my face, but I held it tight.

"Don't touch me," I hissed.

She stopped pulling. "Are you crying?" she asked.

"No," I lied as I tried to dry my eyes and nose under the blanket.

She took no notice of my answer. "Why are you crying?" she said.

I said nothing.

She must have sensed that she made me feel better just by being there—that I had stopped crying—and although I was still hiding under the blanket, she began to tell me about her parents and her three sisters, who lived in a village in the west of Afghanistan. She had had no news from them for several weeks. I felt ashamed of my weakness, ashamed that I had made Sajeda come out of her bed to console me. I remembered what Hameda had told me, that all the girls were homesick for the families they had left in Afghanistan, in faraway provinces. I at least had Grandmother in the same town with me.

"Do you want to go back to your home?" Sajeda asked me.

Slowly I pulled the blanket down, but only enough to show my eyes. "No, I don't want to go home," I said.

"Now, stop worrying. Tomorrow we will be together. Maybe the studies will be a bit difficult for you at first, but we'll all help you. I'm sure you'll like it here just as we do. We are lucky to be here."

Despite her kindness to me, I slept badly that night. I saw a teacher enter the dormitory during the night and wondered what she was doing. I looked at her through my sheet, but I pretended to be asleep. I was afraid she might start talking to me and I had nothing to say to her. Later I found out that three teachers took turns through the night, walking around the school for two hours at a time, comforting any girl who was having trouble sleeping as well as checking the yard and the street outside.

Early the next morning, when I had fallen asleep at last, I was woken up by the *azzan*. A few of the girls climbed out of their beds and prayed on the carpets before going back to sleep. I had never seen children who were that strict about praying. They seemed to care about religion as much as Grandmother did.

But most of the girls slept soundly, and some of them had to be shaken or have water splashed on their faces before they got up and joined the long queue for the bathrooms, where we washed over the sink with a metal can full of water.

Chapter Eight

AS I HAD ALWAYS preferred the company of adults to that of children, the first days at the school were not easy. But the daily timetable was very precise and helped me get used to the routine. I usually hated rules but I obeyed the school's rules like a robot and did what everyone else did.

At first I liked the idea of having a uniform of my own. It made me feel like a real schoolgirl. I would wake up early to look at it, and put it on carefully. It was light blue, with a white collar and white cuffs. But after a few days I got tired of having to change clothes so often during the day. We would wear our uniforms during the morning classes, and we had to change into our own clothes before lunch. Then we did our homework, and after that we had to change again, this time into our white sports clothes for a couple of hours' exercise. Before dinner we had to change one more time, this time back into our own clothes.

At the end of the first week, Hameda summoned me. "We want to give you a chance to decide for yourself," she said. "Do you want to stay in the school or go and live with your grandmother?"

I did not know what to say. It was tempting to say yes, I did miss Grandmother, and yes, I did want to leave. But I was too proud to say so. I had been away from Grandmother for only a few days. The other girls had spent months, even years, away from their families.

"I will stay," I answered.

She smiled. "I knew you would."

Slowly I began to make friends among the girls, who were between seven and sixteen years old. Often the classes would be made up of girls of very different ages, and the younger ones would mock the older ones. "You're too old to be studying with us," they would say, and laugh. They soon discovered that I hated lizards, and when they found one in the yard they would come running to me, pushing it in my face until I screamed and fled.

The girls all looked very different from one another, and they spoke with accents I had never heard before. One girl was from the mountainous Nooristan region, one of the most backward in Afghanistan. When she first arrived she couldn't get used to all the rules and would argue with the teachers. One day she called another girl over, telling her that she had something that would make her ear beautiful. And she pushed a tiny stone into the girl's ear, so deep that the girl had to be taken to the Malalai Hospital, which was run by RAWA.

Sometimes a new girl would bring lice to the school with her, and classes would suddenly come to a stop as all sixty of us were made to sit in the sun out in the yard, while the teachers bent over us and went through our hair. Grandmother had always warned me about lice, telling me how disgusting they were, and I had never had any. I wanted to run away, but the teachers insisted that I stay and be inspected like the other girls. I blushed with shame when I felt the teacher's fingers picking at my hair.

Saima was one year older than me and from a rural family near Kunduz. Her hair was very long, and she wore oil in it all the time, which seemed strange to me. From the beginning she helped me with my homework, especially with the Pashto classes, since she spoke the language much better than I did. We became close, and I admired her because after meeting a RAWA member, she had come to the school, even though her backward family had wanted her to stay at home. When she told me her story, I felt very small because I had done nothing to resist anyone.

THE FIRST FRIDAY after my arrival at the school, the day after Hameda had asked if I wanted to leave, I went to see Grandmother. I was taken by a driver to her house, which was half an hour away from the school. I felt safe when I hugged her. She still had the same warm smell of talcum powder. I also greeted Mujda, but not for long because she seemed less interesting to me.

As I ate the rice and fatty lamb that Grandmother had prepared for me, using a lot of oil the way I liked it, I found her looking even sadder than when we were in Kabul. Even her hair looked grayer to me, and I noticed that the patches of gray had spread from her temples to the back of her head.

I did not have the heart to tell her how much I missed her. When she asked me whether I was happy at the school, I said yes. Instead of complaining to her, I found myself trying to comfort her.

She said her life had changed, that she felt very much older. She told me that she shared the house, which was smaller than ours in Kabul and had just two rooms, with a RAWA member.

I tried to make her laugh. "Grandmother, admit it, I am a burden to you. You have put me in the school because you want to get rid of me," I joked.

She gave me a small smile. "Yes, that's right, you are a real burden, very big and very heavy. Now be gone, or the teachers will tell you off."

I saw less and less of her in the months and years that followed. She always cooked something substantial for me because she knew that the school could not afford to feed us meat or fruit more than once a week. I always brought her a little present from the bazaar, but she often complained that I was never with her, that I no longer loved her.

I would tell her that I was wrapped up in my studies, that I was anxious to catch up with the other girls after all the time I had lost in Kabul, and she would nod and say, "If you are happy, I am happy." I think that even before I started going to the school, she realized that we would never again be as close as we had been in Kabul.

History was one of my favorite subjects. I wanted to find out more about the kings Grandmother had introduced to me. The books I had read in Kabul had taught me that the duty of an Afghan was to obey first Allah and then the king, that the king was second in authority only to Allah. But at the school I was taught that this was wrong—the king was not a second god but an ordinary person, and if he did something that was wrong for our country, it was our duty to disobey him. When I found this out, I asked myself, Why, then, did the kings have such big palaces just for themselves? Surely they could live just as well in a normal house, and spend the money they saved on their subjects.

At the first Persian classes I went to, I felt that I was back on Father's knee. Just as he had given me words to write about, so the teacher would write a word on the blackboard and ask us to think up a good sentence about it—words like *school, Russian, puppet regime,* and *freedom.* I was always quick to come up with something, thanks to Father.

I found English difficult, it was so unlike any other subject I was studying, but the teachers told us it would help us to tell people in other countries about what was happening in

Afghanistan and to study books that had not been translated into Persian. Mathematics, cooking, and sewing were the subjects I disliked the most. During math exams, when the teacher was not looking, Saima was my partner in crime. It was enough for her to catch the desperation in my eyes, when my head was full of numbers that made no sense, and she would quickly pass me her answers for me to copy. Perhaps the teachers realized what was going on, but because I was so much behind the other girls, and good at other subjects, they never punished me. All I got was low marks in math. When there was any sewing to do, I would ask one of my friends to do it for me, and hand it to the teacher, saying it was my work.

My forgetfulness sometimes played dirty tricks on me. I couldn't get used to the highly organized schedule of the school, and while the other girls would write down the dates of exams in their timetables a month early and prepare properly, I would arrive at a class thinking the exam was in one subject only to find that it was in another. I remember once thinking an exam would be Persian. I was good at Persian, so I knew there would be no problem. But when the teacher handed out the questions I sat in cold shock. It was not Persian but math. And because I had been so confident, I had sat on my own and not next to someone who could help me out. So I just wrote my name on the paper and handed it back without any answers at all. When I told the other girls what I had done, they couldn't believe it. "You must be the only girl in the school who doesn't even know what the exam is about," they mocked. I felt ashamed. I tried to remember to ask the other girls ahead of time, but I was often caught out.

No religious faith was imposed on us. The teachers taught us that religion was a personal matter between us and our God, if we believed in one, and they said they would not interfere.

I found out how babies were born. In biology class, we were shown drawings of the human body. The teachers answered all our questions and taught us how to stop having children. They told us that the poor families who had as many as eight children, or even more, were putting at risk the health of both the children and the parents. Parents should have big families only if they could afford proper food, a good education, and medical care.

The teacher taught us that women must not be the sexual slaves of men, that women were entitled to physical pleasure just as much as men were. Sexual violence against a young woman, we were told, could rob her of feeling pleasure throughout her life. It was the first time I had heard such subjects discussed so naturally, the first time I heard the word *orgasm*.

All of us girls wanted to fall in love one day and to be able to choose the man we would marry. My family's teaching had always been that I would not have a husband imposed on me in an arranged marriage.

But I never thought that I must, must get married. Many Afghans think that this must be a girl's ambition: to grow up in her home—as the old women say, "A girl is a guest in the house of her parents"—and then leave for her husband's house and raise her own children. That seemed small-minded to me. I thought I would never tolerate being confined to the home. I would never be a servant to my husband and be told what to cook and what I must spend and how.

I never thought that I would have a sexual relationship with a man before getting married. We girls learned that this kind of relationship was considered normal in the West, but for me and for all the other girls it wasn't normal, and we didn't like the idea. I wouldn't kiss a man before marriage, nor would I go out alone with a man for an evening if we were not engaged. I hoped I would have what my parents had: they had married, and they had loved each other.

Chapter Nine

UNLIKE THE TEACHER I had for such a short time in Kabul, the women at the school seemed to live for their work. I sensed that Hameda especially would have been bitterly disappointed if I had wasted an hour of class without learning something. She was about thirty years old, not married, and her brother had been killed fighting the Russians. I had great respect for her.

One morning, class was suddenly interrupted and all the girls in the school were called out into the courtyard. Hameda lined us up in rows and stood before us looking solemn.

"Some money has gone missing from the teachers' room," she announced. "It is not the first time this has happened, but this time it has got to stop. I know who did it, but I want that girl to come forward now and tell me that she did it. I will wait here until she comes forward."

No one spoke. I had no idea who had stolen the money, and I wondered why Hameda, if she knew who had stolen the money, wanted the girl to come out in front of everyone else.

After what felt like a very long time, Hameda walked up to a tree in the courtyard, reached up with her long arms, and snapped off a thin branch. "All right," she said, "if the culprit refuses to come forward, then the whole school will be punished."

Saima and I looked at each other, struck dumb. Even a group of teachers who were standing to one side looked at one another in surprise. No one apparently had expected Hameda to say such a thing.

One by one, Hameda called out each girl's name. When my name was called, I walked up to her as briskly as I could. I avoided her eyes as I stretched out my hand, palm upward. The switch hit me hard, but not hard enough to draw blood. I did not feel pain so much as anger that Hameda had punished so many innocent girls when only one was guilty. I could not understand what she had done.

A few days later, another assembly was called. I wondered what we would be punished for this time, but I soon realized that it was Hameda who was being put on the spot. The assembly was called "the complaints session," and my friends explained to me that we girls were encouraged to make any criticism we wanted of the school and of the teachers. To me it seemed a crazy idea. Surely everything we ever said to the teachers should show our respect for them?

At the assembly, several teachers stood up to say that Hameda had done wrong. She had let her desperation cloud her thinking, and she had risked losing authority. I was embarrassed for her and stayed silent when the teachers asked what we all thought about the incident. In any case I was too shy to speak in front of so many people.

But Saima did stand up. She said Hameda's attitude had been a mistake, and that it would have been much better to find the guilty girl and punish only her.

I expected Saima to be admonished or caned again for daring to say such a thing. But instead Hameda stood up, admitted she had been in the wrong, and apologized to us all. I had never seen an adult apologize before. Grandmother and my parents had never apologized to me for anything.

Later Hameda explained to me that there was nothing strange about children criticizing adults. "That is how a democracy works," she told me. "Everyone has to be free to say what they think." The complaints sessions were held once a month, but I never spoke at them. Several teachers even criticized me for my lack of participation. "You mustn't think that you are getting yourself good marks by staying silent," they said. "It is your duty to criticize people in authority if you believe in your heart that they are doing something wrong." But I was stubborn and stayed silent.

Even the films we were shown were usually about some form of resistance. I saw *Spartacus* perhaps a dozen times, I also loved *Julia*, in which Jane Fonda played a Jewish writer on a mission to smuggle money across Nazi Germany, and *The Fifth Offensive* with Richard Burton. He played Marshal Tito in the film, and I thought that the war he was fighting in Yugoslavia was very similar to the war in my homeland. I had no idea of this at the time, but scenes that showed lovers kissing too passionately were cut before the film was shown to us.

It was the fact that there was always a guard at the school gate that made me realize what was special about my school. I never found out whether he was armed, but the other girls told me that he had been given shooting practice. I started asking myself what he was defending, and gradually I came to understand that he was defending what the school stood for.

The school was funded by RAWA, thanks to donations made to it by supporters both in Afghanistan and elsewhere and to the money it raised through the sale of carpets and handicrafts that were made mainly by women in the Afghan refugee camps in Pakistan. The school defended the ideals that Mother had fought for when I was a child. It was RAWA that chose the teachers, that selected the children, and that decided how the subjects would be taught. What we were being taught there put us in danger of attacks from Afghan fundamentalists living in Pakistan, so much so that we needed to be protected day and night. Quetta and its outskirts were full of Afghan refugees, and many of them were dangerous men.

None of us were ever allowed to go out on our own. Whenever I went to visit Grandmother, I was taken by a RAWA driver, and I was always told to get into the car before we went out of the gate. School outings were rare. Once, when we had a day at Jaheel, a lake in a park in Quetta, where we sat on carpets spread out on the grass to have a lunch of boiled potatoes and boiled meat, we were watched over all the time by men who were supporters of RAWA—often relatives or friends of members.

It was because we could not go out easily that we had two hours of sports, including gymnastics and badminton, every afternoon. The teachers tried to make us spend as much time as possible in the open air. Whenever it was warm enough, we put carpets down in a corner of the courtyard and sat outside to do our homework.

Some of us would invent diseases for ourselves, hoping to be taken to RAWA's Malalai Hospital for a few days. But the teachers usually found out when we were pretending to be sick. Malalai was no ordinary hospital. There were RAWA posters in its wards, and the staff told women patients about their rights and encouraged the illiterate women to learn to read and write.

During the Russian occupation of Afghanistan, Pakistan, hoping to forge an Islamic bloc that would range to Central Asia, had allowed the fundamentalist Mujahideen factions to use its territory as a base of operations. These factions had set up offices and thousands of *madrassa* Islamic schools in the refugee camps and across the country, which meant that hordes of turbaned students wanted to punish RAWA not only as a threat to their values but also as the most outspoken critic of the crimes of the Mujahideen. Entire areas in the Pakistani towns where we were present, including Quetta and Peshawar, had been turned into no-go zones by the most violent Afghan exiles. RAWA was never allowed to put its name on the sign outside the school.

But despite all this, I felt freer at the school than in Kabul, although I never forgot the suffering I had seen in my city. When I wanted to be alone, I would go and sit in the dust in my corner of the courtyard. History and geography lessons especially made me homesick. The teacher would speak about Afghanistan's resistance to foreign invaders throughout the centuries, about its beautiful mountains, and I would remember the stories I'd heard as a child. Once a girl said to the teacher, "But there is nothing in our country today. Why should we love it? Shouldn't we just love Pakistan, where we live?"

I never had such doubts. Afghanistan was mine and I should love it, and if there was nothing there now, then I should help to build something, although I did not know what. I spent many hours sitting in the courtyard thinking about what I had left behind. I imagined a country where people were afraid. But I never loved Pakistan. I never felt that it could replace my homeland.

OF ALL THE CLASSES, it was the ones in which I could find out about RAWA that interested me the most. I was fifteen when I learned more about the association from Soraya, who taught us political studies. She was older than the other teachers, and as kind as Mother had been. I knew little about her except what the older girls always whispered to new ones—that she was very much involved in RAWA's clandestine activities. We knew her as a brave woman whom we must look up to.

From Soraya I learned the meaning of *democracy*, of *human rights*, of *feminism*. I was told that if men were not allowed to become members of RAWA, it was not because we were against men—we needed their help for the organization to continue working—but because of its very nature.

Soraya wanted us to read as much literature as we could, and to read about the two world wars, about Nazism and about fascism. She never referred to herself as our teacher, saying that we had much to teach her. We called her "Sister."

When I asked her, "How long does it take to make a democracy?" she said to me patiently, "There is no magic recipe"—words similar to those Mother had used when I asked her how she could help Afghan women. Soraya never laughed at any of my questions, and she never used a red pen to mark my essays, preferring blue or black because, she said, "I am not here to judge you."

Sometime after I had settled at the school, the distant cousin who had gone to Canada to study and then decided to settle there came to see me. He had found out where I was through RAWA. He told me about his new life and offered to take me back with him and help me continue my studies there. I would learn English, he said, and then study anything I liked.

He was against the idea of resistance. He said it was useless to try to change things in Afghanistan.

I thought of the thousands of girls in the furthest villages of Afghanistan who were much more talented than I was but who would never be offered such an opportunity. I could not imagine leaving Grandmother, or my classmates.

My cousin was shocked when I told him that I wanted to stay where I was, that I could not conceive of living so far from my homeland. When I told him I believed that "to love your country, you must be ready to die for your country," he shook his head in disbelief. "You are only a child," he said. "Where did you get ideas that are so much bigger than you?"

I did not tell him that I thought his heart was as small as that of a bird.

WE GIGGLED when Soraya announced that we must all change our names. "You must know that there are security problems at this school, that we have enemies," Soraya said. "One day you will understand. Soraya, for example, is not my real name. But please continue using it."

For us it was only fun. I was pleased—I thought it meant that perhaps I was becoming important. All the girls at the school, from the age of twelve upward, were told to use false names, so we older girls felt superior to the younger students. We were no longer children now because we had new names.

It took time for me to get used to mine, which Soraya had

chosen for me, writing it on a piece of paper that she gave me. Over the first few days, the teacher would have to call me several times before I finally realized that I was the one who was supposed to go up to the blackboard. Many of us made the same mistake again and again, and the whole class would laugh.

I chose the name Zoya much later. A Russian journalist had come to see us, and I spent some time with her because she wanted to find out about women's rights in Afghanistan. I half expected her to be blond with green eyes, like the woman-soldier who had offered me a chocolate in Kabul, but she was dark-haired with brown eyes, and because she was so interested in our work for women, I quickly forgot that it was her nation that had occupied Afghanistan for so long.

When I accompanied her back to her hotel and we said good-bye at the door, she kissed me and started walking away, but then she stopped and turned back.

"I hope you don't mind," she said, "but I have one last thing to ask you."

I smiled at her and waited.

The Russian writer had tears in her eyes. "I had a daughter. She got sick with cancer and she died. Her name was Zoya. I miss her, and I would like to ask you to take that name for yourself. Nothing would give me more pleasure."

I was moved by her request and did not hesitate. I said that yes, I would take the name Zoya. I did not even think of the Russians who had invaded Afghanistan—I knew there was a huge difference between a country's government and its people.

A few weeks afterward, I found out that Zoya was also the name of a woman who had taken part in the Russian Revolution. Asked by the czar's police whether she knew where Stalin was, she had answered, "He is here, in my heart." The officer answered, "Well, if he is there, that is where we will kill him." He pointed a gun at her heart and fired.

Chapter Ten

I HAD BEEN at the school only a couple of years when Saima and I decided to organize a meeting of a few friends to discuss our future. I was sixteen years old, and I was becoming more and more frustrated with having to go to class every day and worry about math exercises while my country was plunging deeper and deeper into the nightmare of war.

From my teachers, and from the BBC Persian Service, which we listened to on a radio we borrowed from them—making notes as Soraya had taught us—I learned that life had only got worse for those who, unlike me, had no choice but to stay behind. The shelling of Kabul had continued practically without a pause since the Mujahideen had taken power. Over the past year, 1994, rival factions had imposed a food blockade on Kabul, and many in the city were at risk of starving to death.

But there was a new protagonist in the fighting. The failure of the divided government to set up an Islamic state had prompted many former Mujahideen fundamentalists to rally behind Mullah Mohammed Omar, and this new faction had come from nowhere to strip local warlords around Kandahar, a city in the south of Afghanistan, of their weapons and, in November, to take the city.

This was the birth of the Taliban. In their thousands, they went to give thanks to Allah at the shrine of the Cloak of the Prophet Mohammed, which makes Kandahar one of the most sacred places in Afghanistan.

But their success was not a matter of divine intervention. Mullah Omar could count not only on the support of the most fundamentalist forces among the Mujahideen but also on the men from the *madrassa* Islamic schools, which existed both in Afghanistan and in refugee camps in Pakistan. The Taliban should also have thanked the Pakistani authorities, who had given up on the Mujahideen government and decided to back the Taliban instead as a better tool through which to try to wield influence over their neighbor. Pakistan became the Taliban's biggest supplier of arms.

On the December night when we had decided to hold the meeting, the BBC Persian Service reported that the Taliban had seized control of several more provinces as they spread like a cancer through Afghanistan. After switching off the radio, we met in the study room where we usually did our homework. We sat on the carpet huddled under blankets, sipping black tea to keep warm.

"How many years do you plan to spend at school?" Saima asked us. "Do you realize that if we stay here much longer, we will become adults, and we still won't have done anything for our country or for RAWA?"

"What on earth do you think we could do for RAWA? We're only teenagers," another girl said.

"Wrong," I said. "There's a lot we can do. There's a lot of work, and if one thing is certain it's that we're not taking any part in it by sitting on our backsides in class."

We talked late into the night. Saima and I argued that we were old enough to write for RAWA's publications and to take part in the demonstrations that it held in Pakistan. We were so fired up that several of us spilled our tea on the blankets. The meeting ended when I said, "Tomorrow Saima and I are going to speak out, whether everyone here likes it or not."

When we went to find Soraya, we told her that we had very much appreciated our time at the school, but we felt that now was the time for us to contribute to the work of RAWA. She showed no surprise at our request. All she asked was that we think hard and long about what that meant before making up our minds, and then come back to her.

I did not want to lose any more time. "Sister, we have thought as much as we need to. Our minds are made up, and we want to start tomorrow," I told her.

A few days later Soraya called us to her. RAWA had discussed our request and would arrange for us to leave the school and go to live in a safe house, one of several that RAWA had in

Quetta and other Pakistani towns. They were mostly for the association's younger supporters, groups of whom would live in the houses with RAWA members watching over them.

Soraya told us that we would start our work once we had moved to the safe house.

After so many years of waiting, I was finally to start fulfilling the promise I made on the death of my parents. For the first time in my life, I felt the joy of independence. The decision to leave Afghanistan, to go to the RAWA school, had been taken for me by others. But this was my decision, and I had no regrets about leaving the school.

As soon as I could, I rushed to tell Grandmother. "Daughter," she said, "your life is a special life. I want to see you like your mother and your father."

I WAS PROUD of the fact that we girls ran the RAWA safe house. We had our own daily budget, and we divided up the different jobs and drew up a timetable of who would do what and when. None of us could cook properly, and we ate just because we had to. One of the jobs was sentry duty, and at night we organized a series of two-hour shifts for watching over the house. We kept a gun in the house and hoped that we would never have to use it.

I no longer had to wear a uniform and could wear my own clothes, many of which Grandmother made for me. I still had classes, but only in history, political studies, and English, which the teachers from the school came to give us at our house. They rushed us through only the most important subjects on the history and political studies syllabus, gave us a crash course in English, and after a year the classes were finished. That was the end of my schooling.

Apart from the classes, I was free to read anything I wanted, and I raced through the writings of Bertolt Brecht, in Persian translations. More slowly, I read the speeches of Martin Luther King in English. For days I repeated to my friends the maxim of Abraham Lincoln, which I discovered in Brecht: "You can fool all the people some of the time, and some of the people all the time, but you can't fool all of the people all the time."

The first work I did for RAWA was to write articles on Afghan events for *Payam-e-Zan (Women's Message)*, a magazine that the association had started a dozen years earlier. I learned to write as if I had to defend the choice of every word. For years I had believed that the more complicated a word, the more beautiful it was. When I read poetry, I thought the best words were the ones I didn't understand and had to look up in a dictionary. But Soraya taught me to use words that were as simple as possible, partly because many Afghans could barely read and write.

She also taught me that politics was not about long discussions among what she liked to call white-collar politicians; it was about talking to poor, ignorant, and backward people and showing them that they had a future. "Never speak to the poor as if you are a teacher who knows everything," she said. "Never forget that even the most backward peasant can teach you something."

Three months after we had started living in the safe house, Soraya brought us a thick sheaf of papers that had been clipped together. It was a manual, a collection of accounts by RAWA members of their experiences, and Soraya told us that we should read them attentively and profit from them.

When my turn came to read the manual, I was transfixed. So much so that I read it aloud to Saima and the others. I would also read it late at night, using a flashlight under the blanket so as not to disturb the other girls in the room. The accounts were handwritten, the pages well thumbed.

One member recounted how she had been arrested and jailed in Kabul, and was told while she was in prison that her brother had been killed by the authorities. To force her to reveal RAWA's secrets, she was kept awake for several days at a stretch. The guards would beat her every time she closed her eyes. But she did not betray the association and was eventually released from jail.

Another member was arrested at her school, one of the best in Kabul, when the Russians were still occupying it. She was a teacher, and had passed some RAWA publications on to a colleague of hers. The men of the KHAD, the Afghan secret service that was inspired by the KGB, found the documents during a search of the colleague's house and arrested her. The papers were against the puppet regime that did the Russians' work for them at the time, and called them traitors to Afghanistan.

The police released the colleague when she denounced the RAWA member, and arrested her instead—even though she had a three-month-old baby girl. Her baby went to jail with her. The member denied that she belonged to any illegal organization, but they kept her in a prison for a whole year. Today, her daughter is grown-up and is very active in RAWA. I am always joking with her that she was a little young to be a dangerous criminal so fearsome that she must be kept behind bars at three months!

The accounts described the different tortures inflicted by the KHAD—the way they would tie up prisoners and leave them in the hot sun for days on end, pull out their nails one by one, or give electric shocks to their sexual organs.

I felt sorrow over the suffering inflicted on these people who had been members of our association. When I discussed the manual with Soraya, she told me that it was impossible to predict how someone would react under torture. But I swore to myself that whatever happened to me, I would never betray my friends. I could never live with the thought that a friend had died because of me.

It was Soraya who told me in detail about the life of Meena, the poet who had founded RAWA. I had seen her picture at the entrance of the school. She was a student of Islamic law at Kabul University when she created the association at the age of twenty in 1977, and originally its aim was only equality for women. Then the Russians invaded, and RAWA went underground and started fighting against them, but only with nonviolent means.

RAWA did not campaign for any particular party but for a free, democratic Afghanistan.

In her writings, Meena called the women of Afghanistan "sleeping lions," who would become powerful when they finally awoke. One of her poems is called "I'll Never Return":

I'm the woman who has awoken,
I've arisen and become a tempest through the ashes of my
burned children.
I've arisen from the rivulets of my brother's blood,
My nation's wrath has empowered me.
My ruined and burned villages fill me with hatred against the enemy.
Oh, compatriot, no longer regard me as weak and incapable,
My voice has mingled with thousands of arisen women,
My fists are clenched with the fists of thousands of compatriots
To break together all these sufferings, all these fetters of slavery.
I'm the woman who has awoken,
I've found my path and will never return.

Although she went to live in exile in Quetta, where she set up the Malalai Hospital, she knew that she was still in danger. She received several death threats and even told the authorities in Pakistan about them, but the police ignored her and gave her no protection whatsoever.

She was at her home in Quetta when an Afghan agent of the KHAD strangled her with tow. She was thirty years old.

Soraya told me that one of the men suspected of involvement in the murder plot was Gulbuddin Hikmetyar, the warlord who for three years shelled the people of Kabul from the mountains to the south after the Russians had left. He was responsible for the deaths of twenty-five thousand people in the capital. The life Meena led still inspires us.

"You know the dangers," Soraya said to me. "Are you prepared to accept them? You may achieve nothing in the way of money or power. But if you choose this work, you must prepare yourself for the possibility of being arrested and of being tortured to make you reveal what you know. You may lose your private life, and you may also lose your life. Remember that the door is always open if you want to leave. Maybe the time will come when you will be too scared or too tired. At that time you should leave, but you should always keep RAWA's secrets in your heart."

I did not hesitate. "I know that you have made many sacrifices, and I am ready to do the same," I replied.

Chapter Eleven

LIVING AND WORKING in the safe house meant that gradually I became accepted as a member of RAWA. The association did not believe in formal initiation ceremonies, and one day I was simply handed my membership card. Soon afterward, Soraya told us that the Mujahideen warlords had put several members of RAWA on their unofficial death list. Our headquarters in Pakistan received hate mail and threatening phone calls. Several women who said they wanted to become members were turned away because they were suspected of being infiltrators sent by the warlords.

We girls were inexperienced, and we made mistakes. One night soon after we came to the safe house, one of the girls slipped a lizard into a friend's bed while she was sleeping, and woke her up. The girl jumped out of bed and screamed in fright. Soraya warned us that this kind of prank could put our lives in danger because a neighbor might hear us and become curious about what was going on in the house.

Our house was in a poor district, and the neighbors were more friendly and inquisitive than those in the smarter areas of the town. Soon after we moved in, an ugly old woman came to see us. It would have been very rude to leave her out in the street, so I had to invite her in. She started firing questions at us, asking why we were all in the house. I told her we were sisters.

The old woman had a sharp eye. "But you don't look very much alike," she said. She was right. We were all from different tribes and it showed.

"But then again," she continued, "I can see that yes, you do look alike."

She wanted to know more. "Where are your parents? Surely you are not living here all on your own?"

I launched into a complicated story. Our parents were divorced, our father had gone to live in America, and our mother had sent us here to live with a brother. The old woman tut-tutted in disapproval, then left at last.

We had a more serious incident with a police patrol that was called to our house a few weeks later—perhaps by the ugly old woman—when more than a dozen of us had gathered there to record some songs for RAWA to distribute to supporters. Perhaps whoever had alerted the police was surprised to see such a large group of girls talking and laughing on the terrace after the recording. They must have been surprised that there were no older men or women to be seen, and they were certainly shocked to see young men arriving at the house (they played the musical instruments to accompany us) and then staying into the night. This didn't shock RAWA—we were treated as responsible adults, and the only rule was that men and women were expected to sleep in separate rooms if at all possible.

The laughter was inevitable. I think that Afghan people, when they are having fun together, laugh more than any other people in the world. By the time the three police officers arrived, armed and in their dirty gray uniforms, it was midnight and we had gone to sleep on the carpet. We had no choice but to let the officers in, and they started picking their way among the blankets, lifting them up and mumbling in surprise when they found girl after girl, and then a few young men.

We told them that they were all relatives of ours who had come from Afghanistan to visit us, but it was clear that the officers thought we were prostitutes. They threatened to take us all to court. In the end we had to call in some older members of RAWA, who testified that they were our mothers, and we settled the question by paying a steep bribe to the officers.

The police were corrupt, and it was often necessary to pay them to leave us alone. Small problems usually cost us fifty rupees—what the police liked to call "tea money." But the police were always trying to find RAWA's safe houses and to identify the association's members. If they had realized what they'd stumbled across, they would have arrested our male supporters and questioned them without a moment's rest until they found out all they knew. We girls could also have been dragged to prison, which was not safe for young women.

Sometimes I made stupid mistakes that could have cost my friends a great deal. I would forget to arrange for a bodyguard to take a RAWA member to a meeting, and she would be forced to take a risk and get to it on her own. In Quetta it was dangerous even to keep a member waiting in the street for any length of time because the town was not safe. I always apologized, and I was never punished for such mistakes.

But what made Soraya really angry—the angriest I ever saw her—was some pieces of *nan* (unleavened baked bread) that we had thrown away. We liked to eat our bread fresh and warm from the shop, and every day we would throw away the bits left over from the previous day, stuffing them into a plastic bag that lay in the dirt outside the kitchen door.

When she caught sight of the contents of the bag, she called us all to her. She was literally red with anger, but she did not raise her voice. "Look at this. You should be ashamed," she said. "People are dying all around you because they have no bread to eat, and here you are behaving as if you were the daughters of kings. This is an insult to the poor, quite apart from the fact that you are wasting two rupees on fresh bread when you should save everything you have to buy the vitamins and proteins that you need at your age.

"Do you have any idea where the money you are spending comes from?" she continued. "It does not fall from the sky as you seem to believe. It comes from the sweat of our members, from the generosity of our supporters all over the world, who have no idea that you are wasting it."

We all felt ashamed. For the next three days, Soraya forbade us to buy fresh bread. We felt even more ashamed when she sat before us at mealtimes and ate the pieces of stale bread she had made us serve with the other food. We had to follow her example. We struggled to scrape the gray mold off the pieces of bread and to make them less foul by soaking them in water, warming them on the gas fire, and washing them down with tea.

I can still taste that old, moldy bread in my mouth.

I LEARNED TO CHECK whether I was being followed, to know all the streets around my home, to take long detours when I was going to a safe house, never entering if I had any suspicion that someone was behind me, and to look for a possible escape route every time I entered a new building. We knew our phones were tapped and never dared speak freely.

One afternoon, when a male driver was taking me home from a hotel where I had met a foreign journalist to give an interview, I noticed that a car was following us. I guessed it was the ISI, the Pakistani intelligence service, which like the police had the job of monitoring our activities, trying to identify us and locate RAWA's safe houses and offices. The ISI spies had spotted me meeting the journalist in the hotel and probably knew that I was from RAWA.

I asked my driver to stop. The spies pulled up, not very discreetly, a short distance behind us.

I got out of the car and walked up to them. "What can I do for you?" I asked through the open window. The two men stared at me, too surprised to say anything.

"As you no doubt know, I am from RAWA. We have a telephone number, and you can call us if you'd like to meet us. I am not going anywhere interesting, I'm only going to the market. You're not going to find a safe house or any other of our members by following me now. So may I suggest you stop following me?"

I pushed my membership card, which carried a false name, under their noses, but they ignored it and just smiled sheepishly at each other. One of them finally spoke. "Why do you hide your offices and why do you use false names?" he asked me.

"Because Meena, our founder, was murdered here in Pakistan, and because the Pakistani government supports the Afghan fundamentalists," I replied.

They admitted they were from the intelligence service, and drove off.

I WAS GIVEN more tasks to do. With some other members, I was sent to the main bazaar on a busy Friday to distribute our magazine, *Women's Message*. I was watched over by a male supporter in case I was spotted and challenged by the police or the Afghan fundamentalists. Wearing a veil, I crisscrossed the dark alleyways, looking out for Afghans, whose complexions were paler than those of Pakistanis. The sight of Afghan faces in the bazaars always made me feel homesick.

I came across a man who sat in the dirt with a small pile of onions in front of him, and we started talking. He told me that he was from Kabul, that he was an engineer by training but had lost everything. He told me he wanted to buy the magazine, that he wanted to read it aloud to his children. But he did not have even twenty rupees to pay for it.

"Could you wait until I sell some onions?" he asked me. I gave it to him for free.

IN SEPTEMBER 1996, the men who claimed to profess religious purity took over my city. After marching north from Kandahar, the first city they had conquered, where Mullah Omar dared to drape himself in the Cloak of the Prophet in front of his followers, the Taliban won control of Kabul—but not before first firing shells and rockets at the capital, just the way the Mujahideen had done.

The Taliban's first act in Kabul was to drag Mohammed Najibullah, the ex-president and the former head of the KHAD secret service, from a supposedly safe United Nations compound, in the dead of night, to the presidential palace. There Najibullah was castrated, then shot dead. The Taliban hung him up for display in the Ariana square with a noose made of steel, the wire cutting into his bloated flesh. His brother suffered the same fate. The Taliban stuffed banknotes into the mouths and noses of the hanged men and attached more notes to their toes, as a symbol of humiliation.

Day after day the Taliban published decrees that spawned the harshest theocratic state in the world. Women were ordered to wear the *burqa* outside their homes. They were banned from appearing on the balconies of their houses. They could go outside only if they were accompanied at all times by a *mahram*, a close relative. They were banned from working. At certain times during the Ramadan month of fasting, they were simply not allowed on the streets.

Women who were sick could only be treated by women doctors. Girls could not go to school—according to the Taliban, schools were a gateway to Hell, the first step on the road to prostitution. Women were not allowed to laugh, or even to speak loudly, because this risked sexually exciting males. High heels were banned because their sound was also declared provocative. Makeup and nail varnish were banned. Women who failed to respect such edicts would be beaten, whipped, or stoned to death.

The *hammams* were closed. Men were ordered to grow their beards. Music and television were banned, and so were games, including kite flying. What could be more innocent, I asked myself, than a child playing with a kite?

All this, I thought, was the work of a bunch of criminals who didn't even know how to write their own names.

Soon we learned that the Taliban had put all members of

RAWA on a death list. I read articles in the newspaper in which the Taliban leaders called us infidels, CIA spies, prostitutes who wanted to go out in the streets and sleep with men. Whenever they found a RAWA member, the leaders swore, they would execute her immediately without trial because we must all be wiped off the face of the earth. Even if it took all the Moslems of Afghanistan to do it, they would hunt us down to the very last member and eliminate us. The blacklist that we knew existed against us under the Mujahideen warlords had become official under the Taliban.

We knew we could not count on any protection from the Pakistani authorities. Benazir Bhutto, the prime minister, greeted the Taliban conquest of Kabul with the statement that if the Taliban managed to unite Afghanistan, it would be a welcome development.

THE WORSE the situation became in Afghanistan, the more the work of RAWA became vital to my existence. It was the most important part of my life, more than anything or anyone else—more important, even, than Grandmother.

I admired the self-sacrifice that Mariam, a friend of mine, made on the day of her marriage in Islamabad. She came straight back from work in her office clothes, got married, and then sat down with us, and her new mother-in-law, for an ordinary meal of chicken and rice. She had barely started eating when the telephone rang: it was RAWA, asking her to leave immediately and travel to another town, three hours away.

Without hesitating, Mariam said good-bye to her husband, told him she would be back in two days, and left. He did not complain, and Mariam's mother-in-law remained silent, but her expression was frosty. Soon after we sat down again to continue the meal, she began to vent her frustration on us.

"Strange, isn't it," she asked with a cold smile, "that a bride should leave her husband on the evening of the wedding? I think RAWA has some very special traditions that you cannot find anywhere else in the world, in the East or in the West. These traditions are so amazing that I have no idea what philosophy they come from."

I exchanged glances with my RAWA friends. I could see from their eyes that, like me, they wanted to laugh, but we all struggled to keep silent.

A few mouthfuls later, the mother-in-law spoke again. "This evening may not matter very much to you, but it is very important to me," she said.

Again we exchanged glances. Again we kept quiet.

But she was determined to provoke a reaction from us. "It is very good to belong to this organization, but I never thought political commitment meant that a bride could leave her husband on the evening of her wedding without even asking him for permission."

I could stay silent no longer. "You should not say these things," I said. "We have only respect for you, but Mariam has a job to do. She did not leave to go and enjoy herself. Instead of criticizing her, you should be proud of her."

In defending Mariam, I felt I was defending myself, because I hoped that I would have acted in exactly the same way. In fact, I have never had a private life, and I have no regrets about this. I do not see anything beautiful in me that a man could look at in a special way. I have never dreamed of a man looking at me, nor have I fallen in love. Nor do I feel sad that I have never known physical pleasure with a man. It has never been important to me, and I have not had the time to think about it.

Only if one day there is peace in my country, and a democracy in which men respect women, can I think of marriage. It is very important to me that the man I share my life with respects me and what I do. In that, Father is a model for me, because he respected Mother and her work.

PART FOUR

The Silent City

Chapter Twelve

I DISLIKED THE SHOP and the shopkeeper, but above all I hated the garment that I was being forced to buy. I had come to the bazaar to buy a *burqa*, and I spent as little time as I could in the shop that displayed them so proudly in its window as if they were the latest fashion. I thought they looked like disgusting sleeping ghosts. I saw a blue one—the most common color—that looked about the right size for me, and tried it on over my shirt and trousers. It was the cheapest, made of polyester.

"I can't see. I'm going to fall down in the street with this thing on," I complained to the shopkeeper as I struggled with the heavy material. I had had it on for only a short time, but already I was sweating in the June heat.

"Don't worry, you will practice, and after that you will have no problem," he replied.

I wrenched it off, handed the man his five hundred rupees as he folded the *burqa* neatly and put it in a plastic bag, and then I got out of the shop as fast as I could. It made my blood boil to hand over money for something I loathed. If I could have, I would have set fire to the whole shop. But it would have been dangerous for me even to tell the shopkeeper what I thought of his wares. He might have started arguing with me and attracted the attention of a police patrol.

I had no intention of practicing as he had recommended. I knew I would get plenty of chances to get used to it very soon.

It was the summer of 1997, and after almost three years with RAWA I was at last being sent on a mission to my homeland. My task was to see what could be done to help several of our members who had written to RAWA in Pakistan saying that they had problems they wanted to discuss with us. It was too risky for them to describe these problems in their letters, which they smuggled across the border with the help of supporters, and that is how I got my chance.

I was also told to find out whether we could bring women from Afghanistan to take part in a street demonstration that we would soon be holding in Pakistan. The aim was to bring as many as one to two thousand women from Kabul without the Taliban spotting any of them, either on the way out or on the way back.

I was to travel with Abida, a friend and RAWA member who was several years older than I and had been back to Kabul before, and with Javid, a middle-aged male supporter. He would follow us like our shadow and be our *mahram*.

Javid had been growing his beard for weeks in preparation for a trip to Afghanistan. A driver would take us as far as the border. Apart from my travel companions, only six other people knew about the mission.

I hated packing the *burqa* into my travel bag. For the Taliban, who would soon be celebrating the first anniversary of their conquest of Kabul, it would serve to guarantee my dignity and my honor, as I would be obeying the decree that a Moslem woman must observe complete *hejab*, or seclusion from society.

Later, the Taliban did not stop at simply ordering that women wear *burqas*. They dictated that Hindu women should all wear yellow *burqas*. In Afghanistan, yellow is the color of sickness, and of hate, and for the Taliban all the members of the Hindu minority were infidels. The women had to wear yellow just as Jews had to wear the yellow stars imposed on them by the Nazis.

I did not expect to see anything beautiful, or anything that would make me happy, during my visit to Kabul.

KABUL WAS a graveyard. The river that gives the city its name was brown and cold, and rubbish floated slowly down it. When our minibus pulled to a stop at the bus station in the center of the city, I could not stop my tears. It was evening and already dark, and the buildings, so many of them just empty shells, looked like tombs. Despite the devastation before me, I understood when Abida breathed in my ear, "Oh, my lovely Kabul," and I nodded.

As soon as we stepped out of the minibus, the beggars crowded round us, pleading for alms. I had never seen so many beggars—there must have been two dozen of them. Through my tears, I saw one little boy about ten years old who had half his right arm missing. I guessed it had probably been ripped off by a mine that he had picked up, one of the thousands that were the legacy of the fighting since the Russian invasion. As he stretched out his left hand, he sang a song to me, all about potatoes and meat. All he ate was dry, hard bread, he sang, and he had never seen the color of potatoes or meat, never tasted them. It was a sweet melody, but I had no money to give him.

We had taken only a few steps away from the beggars when a woman stopped us. From the way they shuffled toward us, stooped and frail-looking, we guessed they were old women. But we soon realized they were in the pay of the Taliban religious police, the crazily named Amar Bil Maroof Wa Nahi An al-Munkar, the Department for the Promotion of Virtue and the Prevention of Vice.

I had been warned about such women before I left Pakistan. They helped the Taliban keep tight control of everyone on the streets. It was easier for old women than for the Taliban to check on what you were hiding inside the *burqa*.

"Show us your bags," one of the old women commanded.

I felt sick with fright, unable to speak. I thought not only of the RAWA publications in our bags documenting Taliban outrages, but also of the more important letters that were hidden in a pouch tied around my stomach. If the old women found any of our papers, our mission to Kabul would be over before it had even started.

The old women were illiterate and would not be able to understand the documents. But they would know from the photographs in the publications that it was banned material, and report us soon enough. I did not want to think of the fate that would await us in a Taliban prison cell.

I heard Abida at my side speak cheerfully to them. She spoke in Pashto, the language that the old woman had used. "Mother, we have just arrived after a long, long journey," she explained. "We are simple girls, we are exhausted, and my friend is not feeling very well. We have nothing special to show you, only clothes."

Thanks to Abida, the old women lost interest in us, and we walked away slowly although we wanted to break into a run. We found a taxi and set out for the safe house where we would stay for the week. I was not told whom it belonged to, and didn't ask.

When I arrived at the safe house, I had not yet taken off my grimy *burqa* when I heard a shout and felt someone give me a bear hug. The embrace went on and on until I felt myself suffocating. After what seemed like minutes, I was finally released and got a chance to take off the *burqa*. I kicked it aside.

I recognized Zeba, one of our most courageous members. I had met her a couple of times when she visited Pakistan. It was Zeba, I knew, who along with others took great risks by filming some of the worst crimes committed by the Taliban, including their public hangings and executions.

I will never forget the film of a woman in a light blue *burqa*, the same color as mine, kneeling near the goal markings on the pitch of a former soccer stadium while a Taliban in a turban pressed the end of his Kalashnikov against her headband. She tried to get up, but a mullah pushed her down again. And then the shot, which kicked up a small spurt of earth as the bullet hit the ground after going through the *burqa* and through her skull.

The executed woman was the mother of seven children, and the Taliban accused her of killing her husband in a family quarrel. The husband's family forgave her, but the Taliban decided to go ahead with the execution anyway.

Later, another execution was captured on film: in the middle of a big crowd as the Taliban used cranes to hang two men on the edge of a busy street. The men were accused of cooperating with anti-Taliban forces. They were blindfolded, their hands tied behind their backs, and a rope slung around their necks. They died almost instantly when the cranes lifted them off the ground, and they were left to hang there for a whole day, their feet swinging at the height of people's heads.

At the main stadium in Kabul, Zeba filmed a public *qasas*, a religiously sanctioned slitting of throats. A man convicted of killing two people was forced to kneel in the stadium, his eyes covered by a scarf. He was given ten minutes to pray, then his hands were tied behind him with another scarf. Then a brother of one of his victims walked up to him, carrying a knife. The brother drew the knife across the man's throat.

I realized that I could not begin to imagine the dangers that Zeba and others were running. Their lives and mine were as different as the earth and the sky.

"About time! I thought you'd never come," she said, smiling at me. Zeba was in her midthirties, but she had so many wrinkles and her hair had so much gray in it that she looked twenty years older. Yet when I asked her how she was, she answered only "Fine, fine."

I sat up until three o'clock in the morning, drinking countless cups of tea as I talked to Zeba and other RAWA members. Then she suddenly stood up and said, "Right, off to bed. We shouldn't keep you up like this. You're going to need all the sleep you can get. In two days' time there's going to be a public cutting of hands at the stadium. I'd like you to come with me and help me photograph it."

"Bed" meant the carpet. I was too tired to think about what lay in store for me, and fell asleep as soon as I lay down.

WHEN THE NOISE of my RAWA friends making breakfast woke me up a few hours later, at first I thought I was in Pakistan. Then I realized where I was, and I was happy to be back. When I went into the yard to wash my hands at the tap and saw Kabul in the daylight, even the mountains beyond the city—which had seemed so peaceful to me when I was a child—looked sad. But the fact that I had seen them again, after so many changes in my life, made me feel stronger.

Saddest of all, for me, was the fact that there was not even one kite in the sky. The Taliban had stamped out one of the oldest traditions of my country and emptied the sky.

I went back into the house and noticed that most of the windows had been draped with curtains that were black on the side facing out into the street but all different colors on the inside. The Taliban had decreed that houses where women lived must have black curtains always covering the windows so that no one could see them from the outside. But the people I was staying with had insisted on some color.

Again I had to put on the *burqa* before I could leave the house. Because I was not used to walking with it, I grabbed Abida's hand as we walked through the streets with Javid. For a report that we had to write up, I was being taken to meet a woman whose teenage daughter had been raped by a Taliban commander in the street. A RAWA member knew one of her relatives.

We had walked not very far from the house when I heard a whistling sound very close to me and, a fraction of a second later, felt a sting on my hand. I thought I had been bitten by a snake, but when I turned I saw a Taliban with a lash in his hand.

"Prostitute!" he shouted at me, the spittle spraying his greasy beard. "Cover yourself and go from here! Go to your house!" He wore a black turban, and I thought his stare was strange until I saw that he was wearing *surma*, a thick black eyeliner, to make himself look more aggressive.

Abida apologized for me and quickly pulled me away. She told me that my hand must have come out from underneath the *burqa* while we were walking. "Please be as careful as you can," she said. "We can't afford to draw any attention to ourselves."

When we knocked at the door of the woman whose daughter had been raped, I told her that we were from RAWA and wanted to help her, at the very least with some words of comfort. She was small and weak, and the strength of her reaction took me completely by surprise.

"If you are from RAWA, you'd better leave right now," the woman snapped.

"Why? We only want to help you," I insisted.

"You say you are fighting for democracy and for women's rights, but your methods are completely wrong. If you have a gun to give me, then you can come in. That's all I need, a gun. I know who raped my daughter. He is a very powerful commander, and I have no other way of avenging her."

I thought for a moment that if only I could have taken off my *burqa*, she could have seen my eyes and understood how much I wanted to help her. But she was in such pain that I could find nothing to say. I was not angry with her. I felt only pity, and I was sorry that I could do nothing. I hoped that with time she would agree to talk to us.

Thousands of women had suffered the same fate as her daughter. In the areas of central Afghanistan populated by the Hazara tribe, the Taliban kidnapped young women as *kaniz*, or servants, and then gave them to their soldiers to marry. Theirs was a strange creed: they could rape women and force them to marry, but they stoned to death women suspected of adultery.

No tribe suffered at the hands of the Taliban as much as the Hazaras. A few months before my journey to Kabul, in September 1997, the Taliban carried out a massacre of Hazaras in the village of Qezelabad in the north of Afghanistan: An eight-year-old boy was decapitated, and two twelve-year-olds had their arms and hands broken with stones as soldiers held them.

I WAS TO FEEL frustrated several times in Kabul. When I visited a hospital with Abida and Javid to determine how risky it would be for us to take pictures there, I saw dozens of patients, young and old, lying on the filthy concrete floor of the corridors with no one paying any attention to them.

Many of the children showed clear signs of malnourishment, the skin taut over their faces and their arms as thin as sticks. I had read that parents sold their children in the street because they could not afford to feed them, or simply gave them away to anyone who could offer them a better life. The toilets were in an awful state—there was urine and excrement all over the floor. There were more Taliban than doctors in the hospital. We saw them patrolling the corridors in their black turbans, lash in hand, picking their way among the sick people.

The women suffered more than the men, because the Taliban would not allow them to be treated by male doctors.

For the Taliban, if a woman was sick, it was better for her to die than to be treated by a man. If she refused to let a male doctor touch her, she would be certain of going to Heaven. If she let herself be treated by him, she would be condemned to Hell. Of course, there is nothing in the Koran to justify this belief.

The only woman I was able to speak to in the hospital told me she could not afford medicine because she was not allowed to work, and had been waiting for days to see a doctor. She was forced to wait so long because there were not enough women doctors left in Kabul. Many doctors, both men and women, had abandoned Afghanistan under the Russians to make a better life elsewhere, in Pakistan, Iran, or in the West, and many more had fled under the Mujahideen and the Taliban. There were no new women doctors to take their place, since the Taliban prohibited women from studying medicine as well as everything else.

In Pakistan I had met a woman surgeon who had left Kabul. She told me that under the Mujahideen, she had been forced to operate by candlelight because there was no electricity when the city was being bombed, and that her shifts lasted as long as twenty-four hours at a time. She had been pregnant, and she lost her child because she had spent so many hours standing in the operating theater. Despite this sacrifice, she felt ashamed that she had left Afghanistan.

Abida pretended she had kidney problems, and managed to talk to one of the few women nurses who were working there. But she was able to ask only a few questions about the hospital before the nurse became suspicious. Taking photographs in the hospital would be a very dangerous task.

A few days later I found out that the Taliban did not hesitate to use their whips against the sick, even in the hospital. I was out walking in the street when I saw a woman sitting in the middle of a busy road, surrounded by a small crowd. She had tried to commit suicide by throwing herself into the middle of the traffic. "Let me die, let me die," she said again and again. She was lucky that there were no Taliban nearby, as they would surely have beaten her right there in the street.

Later, when I managed to speak to her in a quiet place, she told me that her mother suffered from asthma and had gone to hospital for treatment. Soon after reaching the hospital, she had suffered an asthma attack and had taken off her *burqa* as she fought for breath in the ward. A Taliban had burst into the ward and given her mother forty lashes while the daughter watched, helpless to intervene. The nurses had done nothing to stop the beating.

The daughter, who was twenty years old, explained why she had wanted to commit suicide: "If I can't even help my mother when she is sick, then what is the good of living?" she asked me.

I thought of Grandmother, the asthma attacks she suffered, and of how I would have reacted if I had been in the girl's place. It was one of the worst moments of my visit to Kabul, and for a time I felt discouraged. The *burqa* not only killed women mentally, it could also help to kill them physically.

I noticed more signs of the toll that war and the Taliban had taken on people's mental condition. As I walked in the streets, I often saw people behaving strangely. Some men walked around aimlessly, a glazed expression on their faces. I saw one talking loudly and endlessly to himself, without stopping for breath. Another burst out laughing like a madman as I passed. There were certainly no doctors qualified to deal with this.

In the streets of Kabul, the chants of the beggar boys were the only music I heard. As a child, I was used to loud music coming from the shops and the cars. Now, the only cassettes that people were allowed to play in their cars were religious chants with no music, which go on and on, just one voice that has no melody to it at all.

That was also the dominant, hypnotic sound coming from Voice of Shari'a, the Taliban radio station, on which, of course, no woman was ever allowed to speak. The only exception that I heard was a phone-in program, a daring innovation in Taliban broadcasting history, during which male listeners could question a panel of mullahs exclusively on religious matters.

The program did not go quite as planned. One insistent listener who said he was from a small village kept asking how he could decide which was the senior of the two mullahs in his village. The answer "The one who knows the Koran best" did not satisfy him. Neither did the answer "The one who is a better Moslem." He kept asking how he could tell the mullahs apart, until one of the so-called experts told him that he should find out which one had the most beautiful wife. Then the listener was cut off. Never in my life have I heard such a ridiculous and absurd program.

Zeba told me that the only time she could listen to her music tapes was before going to sleep, and she would keep the volume as low as possible out of fear that the neighbors would inform on her if they heard the offending sound.

Before, there were pictures of the most famous singers in all the shops, but now photographs of any kind were banned. So was television. Still, in several of the houses I visited I saw that people not only had illegal television sets, they also had home-made satellite dishes in the yard to catch foreign channels. They would throw a sheet over the dish if anyone knocked at the door. They usually managed to catch only one Pakistani or Indian channel, but that opened up a new world for them. The most technically accomplished even managed to tap into CNN and BBC World Service. For free of course.

The Taliban regularly raided houses that they believed harbored televisions. But they were not always as strict as they were supposed to be. One family who had been caught watching a video of an Indian film were pulled out of their home and given a public lashing in the street. The Taliban shouted at them that it was anti-Islamic to watch such films. Then the Taliban left them outside and reentered the house. When the family dared to return, they found the Taliban sitting around the television set watching and commenting on the film, which was still playing. The Taliban took a bribe from the family and did not arrest them.

Chapter Thirteen

ONE MORNING Zeba came to fetch me at the safe house. "Come, I need you," she said. "We have to photograph the cutting of hands."

The previous day, Voice of Shari'a had announced that a thief would have his hands cut off at the main Ghazi Stadium and had urged all the people to go and watch. Together with several RAWA members from Kabul, we drove to what had once been the football stadium until the Taliban banned all sports.

We were driven in our own cars. The Taliban ordered that men and women travel in separate buses, but if we had done that we would have been sure to lose one another. My RAWA friends had told me about a husband and wife who were separated on the way to a wedding. After a couple of hours they finally found each other and began to argue in front of everyone in the street.

Then they saw the comic side of their misfortune and decided, because it was so late, to forget the wedding and go home to bed. In this country, my friends told me, you're not even free to go to a wedding.

I smiled, but I didn't have the heart to laugh. I couldn't see the funny side of stories about life as it was then in Kabul. I understood that my friends felt differently because they were stuck there. They needed to laugh. But I could only think that for many war widows the rule that they could not go out without a *mahram* was a tragedy. It meant that they could not leave their houses and had no way of earning a living apart from begging in the streets and risking a lashing from the Taliban, or turning to prostitution.

Near the stadium, we saw their patrols ordering shopkeepers to close down and go watch the ritual. I was surprised to see women taking their children with them, but Zeba explained, "They want their children to realize what will happen to them if they ever steal anything. They think scaring them is a good way to educate them."

When we reached the stadium, several thousand people had already arrived and were waiting quietly. We headed for the women's section, which was across the stadium from the part reserved for men. The goalposts were still standing. I had been told that sometimes the Taliban would hang someone from them.

A convoy of a dozen jeeps sped onto the field, and men in turbans tumbled out of them, some holding guns. One man was led to the center, and made to lie down on his stomach, his arms spread out in the shape of a cross. I counted no fewer than five Taliban holding the prisoner down. One of them tied his feet together, while another grabbed his hair and forced his head up off the ground.

A mullah spoke to the crowd, using a microphone. He talked about sin and the Day of Judgment. "This man deserves the punishment that he is about to get," he said. "All those who steal will be punished in this way."

As he spoke, a group of people who must have been the prisoner's family begged for mercy, only to be lashed by the Taliban.

I saw a figure standing to one side, a white scarf wrapped around his face so that it left only his eyes free. "He's a doctor," Zeba whispered to me. "He's frightened that if people recognize him afterward, they will kill him for helping the Taliban."

Together with the other RAWA members, we pressed close around Zeba to hide her from the crowd, and she started taking photographs with a small hidden camera. She was careful not to waste pictures because she could not risk stopping to put in a new film, which would have drawn attention to us.

The Taliban in the black turban drew a knife, knelt down on one knee at the prisoner's side, and started sawing at the man's right wrist.

The blood spurted onto a patch of earth.

I could watch no more. I suddenly felt a pain in my wrist as if I had felt the blade against my skin. I felt faint and sat down on the floor in the middle of the crowd, which seemed to have become frozen.

A few of the women around me cried out against the Taliban and against the doctor, who was binding the man's wrists to try to stop the flow of blood. "One day you will all be lying in that man's place," I overheard one woman say. "May Allah grant that you will be next," said another. But they did not speak very loudly.

The children around me laughed and clapped. For them what they were seeing was entertainment, as normal as the soccer games they used to watch on television before the Taliban took power. And what's more, it was free. I tried to imagine the future of these children. They would all become heartless criminals if things went on like this.

I thought how strange it was that these children were free to laugh at such a spectacle but that I as a woman was forbidden to laugh in a public place.

Sometime later, after another cutting of hands, Zeba took a picture of a boy, grinning from ear to ear, as he held up a man's severed hands, which he had taken from the tree where the Taliban had hung them. The boy was playing with his friend. They had been throwing their trophies to each other across the street and laughed.

I still have that photograph. I admire Zeba for having the idea of pointing her camera at this boy: she had found someone who was so proud and happy to have his picture taken that he would never have denounced her. And I pity the boy for feeling important because Zeba was photographing him, and for failing to understand the horror of what he was doing.

I had felt powerless in the stadium. I had wanted to go to the man and help him, but there was nothing I could do.

THE WOMEN had spoken quietly in the stadium, but at least they had dared to voice criticism of the Taliban.

The strongest challenge to the Taliban's authority I saw was made by a woman I came across when I was walking through a market in the center of Kabul accompanied by a male RAWA supporter who was pretending to be my *mahram* escort. She was standing in front of a vegetable stall and handing over some money to pay for her purchases. I saw a Taliban patrol approach in a jeep with a white flag flying.

The arrival of these patrols often caused a wave of panic. Women who were without a *mahram* would turn to complete strangers and offer to pay them to pretend that they were together. "Please, be my brother," the women would plead. It was a dangerous solution. Women who were found out were whipped, as was the pretend brother.

One of the Taliban—he looked to be little more than a teenager—jumped down from the jeep, marched up to her with his whip, and lashed out at her arm. The woman, who had not seen the patrol, had broken the law that women must not have any direct contact with shopkeepers and must have a *mahram* to buy things for them.

Far from cowering away, the woman turned on the Taliban like a Fury. "I am old enough to be your mother, and you whip me? Don't you feel ashamed?" she shouted at him.

She was so incensed that she even dared to pull off her *burqa* and throw it at the feet of the Taliban. "Here, why don't you wear it yourself?" she mocked him.

She was tall and strong. I guessed she was in her forties. Her attacker was so surprised that he did not know how to respond. No one had trained him for that kind of resistance. All he had been taught was how to whip women. He slunk away. After her victory, the woman retrieved her *burqa*, put it back on again, and continued her shopping. I marveled at her bravery.

Other, more discreet signs of rebellion warmed my heart during my visit to Kabul. As small as they were, they showed me that the people were still alive.

Despite the Taliban attempt to crush it to dust, many women clung to their femininity. Several of the young women I met wore makeup or perfume under the *burqa*, and they visited beauty salons that operated in secrecy. The salons were popular especially with brides, who wanted to make themselves as beautiful as possible, even under the Taliban. Strangely, cosmetics were sold in the shops, although their use was forbidden.

Even wearing something as petty as nail polish, which was banned by the Taliban, could bring terrible punishment. I was shocked to see the young daughter of a RAWA member painting her long nails a bright pink color.

"But isn't it dangerous?" I stammered.

"What am I supposed to do? Stop living because of them? If they want to beat me, let them beat me," she replied.

I was amazed. I knew that the Taliban had cut off the fingertips of some women they had caught wearing nail polish.

I SAW SEVERAL EXAMPLES of dedication in Kabul, and one of those that impressed me most was that of Khalida, a teacher in charge of teaching clandestine classes for some three hundred children in various areas of Kabul. Under the Taliban, girls could not go to school, and boys could study only the Koran, so RAWA had set up these classes for children whose parents were ready to take a risk for the sake of their offspring's future.

Already the Taliban had found out about Khalida's classes from their spies and had told her she must stop teaching. She had told them that she would stop. Then she had simply moved the classes to another safe house. She risked execution if she was caught again.

I went to find Khalida at a small two-room mud house that belonged to a RAWA member and her husband. The couple were part of the security cover—if the Taliban raided the house, they could always say that the children were theirs. Often couples like these would have to move to a new house every five months or so, simply to ensure that the children could study in safety.

I gave the password that had been agreed upon previously, and the couple let me in. Having to wear *burqas* had made passwords more necessary than ever, because you could never tell who your visitors were just by looking through the window as they arrived at your door.

Khalida was teaching a Persian class of only four children when I visited her. In the Kartayi Parwan area to the west of the old city it was too dangerous to hold classes that were any bigger. The children were aged from eight to fourteen and were sitting on the carpet. They had been told to say, if asked, that they had come to visit their aunt, never their teacher. The parents brought each child to the house at a different time and never told them what RAWA was, let alone that it had anything to do with the class.

It was impossible to give them any homework in case they were found carrying it. The children of Afghanistan were allowed to carry a Kalashnikov but not homework.

On the blackboard that rested against one of the damp and dirty walls, Khalida had taken the precaution of writing "I begin with the name of Allah," because Allah was the first word that boys learned in the Taliban schools. Underneath, a child had written "I love my country."

Khalida had placed an open copy of the Koran prominently in front of her. After the class had ended, she explained to me that she always had one out so that if any Taliban burst in she could slip the Persian or math book they were really using underneath it and pretend that the children were studying the Koran, which was tolerated.

Khalida was exhausted and besieged me with her list of demands. "We need a bigger house so that we can teach more

156

children. You know what the children say to me? 'We are so hungry and our stomachs are making so much noise that we can't understand the lesson.' I have to stop the class and go fetch some bread. Can't RAWA pay for their food and their clothes? And I need money to buy the stationery for the children because their parents can't afford it. You have no idea how many children are not coming to me simply because their parents can't pay for the pens and the paper. Can't you pay for that too?"

She was so desperate she started shouting and slapping the carpet with the flat of her hand. I told her to keep her voice down, that we risked drawing the attention of the neighbors. She calmed down.

I felt sorry for her. "I'll ask and see what they say. But you have to understand that we're not talking about free stationery just for your children. It would have to be free for all our classes in Afghanistan," I said.

Khalida's children were not even a drop in the ocean, but they were the future of Afghanistan. Later I was able to send her good news—RAWA could not afford to pay for food and clothing, but it did agree to pay for the stationery of the children who went to its classes all over Afghanistan, and we found her two new safe houses too.

Chapter Fourteen

I NEVER DID see my house. Nor did I see any of the people I had known in Kabul as a child. Once, when Abida, Javid, and I were on our way to some meeting, I saw that we were close to my neighborhood, and I asked the driver to take the main road from which I could see my street. I asked him to go very slowly as we got close to the street.

I pressed my face to the window, the mesh of the *burqa* obscuring my vision. Many of the buildings in the road had been bombed, and most of the shops that were still left standing had closed. I saw that there were no children playing, no hens or goats, in my street. I could not make out the blue door of my old house.

The driver asked me whether he should stop the car for me to get out. He had guessed. I was tempted, but I said no. I didn't want to see it now. In any case, I hadn't asked Grandmother for the keys. Perhaps one day, I thought, if peace returns, I will come back to see my house.

When I said good-bye to Zeba before leaving Kabul, she smiled at me and said, "It's good that we saw each other, because you may not see me again."

I tried to laugh, but something stuck in my throat. "Don't say such things. Of course I will see you again," I replied.

"Well, it's true. You should prepare yourself so that if I am arrested, you are ready for it," she said.

This time, we embraced before I put on the *burqa*. I knew she was right, and I thought about her and the danger of her work for much of the journey. I felt so sick, so tired, and so sad when we crossed the border back into Pakistan that I did not bother to take off the *burqa*. I took it off only once I had reached my home.

When I looked at myself in the mirror, I saw something that looked like the mark of a cage in the middle of my forehead. I realized that when I had slept during the last part of the trip, the mesh of the *burqa* had ridden up over my face and left its imprint on my skin. It was the only mark that my journey to Kabul left on my body, but my heart was wounded.

I RETURNED SEVERAL TIMES to Kabul over the next few years. Increasingly I was struck by the absurdity of life under the Taliban. Once, as I walked in the street, a man started asking me what kind of vegetables he should buy. I thought he was crazy, but then I realized that he had mistaken me for his wife. The *burqa* I was wearing was the same color as hers, and she had stopped to look at something.

Even something as mundane as eating ice cream became a ridiculous undertaking. Only certain shops agreed to sell ice cream to women, because the owners worried that a Taliban patrol would come and beat them for allowing women to gather together. Friends told me that there were no chairs for them to sit on, and they would have to stand there, lifting the *burqa*s off their faces with one hand and trying to eat the ice cream with their other hand under the cloth. They said they looked like clumsy birds with long beaks. They complained about the Taliban while they tried to eat as fast as they could because the ice cream melted fast and dirtied their *burqa*s. Washing them is never easy because all the creases have to be ironed, and women would often only clean the part in front of the mouth.

On one of my visits to Kabul, the film *Titanic* was all the rage, so much so that a man's haircut was called after it. Videocassettes had been smuggled into the city, and boys would go to the barber and ask for the "Titanic cut" inspired by Leonardo DiCaprio. But a Taliban mullah preached against DiCaprio and Kate Winslet, saying they had sinned against Islam by having physical relations before getting married. He decreed that the *Titanic* had sunk because Allah was angry at the lovers' behavior. The Taliban thought the iceberg was fitting punishment. But they had more in store for the stars: if Di-Caprio and Winslet ever set foot in their country, they would be stoned to death.

In the meantime, the Taliban decided to punish any boy who had sinned by getting himself a Titanic cut. When they caught an offender in the street, they would whistle to him as if they were calling a dog. Then they would jeer at him. "Hey, handsome boy," they would say as they pulled roughly at his hair. "What is this hairstyle? We like it so much. Oh, so you want to be an actor in the infidel film?"

Then someone would fetch a pair of scissors and the Taliban would start chopping at the boy's hair. They would send the boy home, his hair a jagged mess. He was lucky if he got away without a whipping.

But the Taliban disapproval did not stop stallholders at markets in the center of Kabul from shouting "Get your Titanic apples here!" or "Titanic cabbages for sale!" One market that moved for a time during the summer to the bed of the Kabul River, when there was little water in it, called itself the Titanic Bazaar. Nothing the Taliban did could quench the thirst of the people in a country where forced marriages were the rule for a story of undying love.

The absurdity of the Taliban had no limits. When the Monica Lewinsky scandal broke in America, one of the mullahs at a Kabul mosque could not even get her name right in his condemnations of her anti-Islamic behavior. He kept calling her "Monica Whisky." She had committed a double sin: not only had she behaved impurely with the president of the United States, but her name, that of a forbidden drink, was also a challenge to Islam.

I never told Grandmother about my journeys to Kabul before I set out. I mentioned them only when I got back. When I told her the news after my first journey, her disbelief gave way to anger and then to relief. "You did well not to tell me before you went," she said. "Otherwise I would never have let you go."

My doll Mujda was living with Grandmother. She was pretty much the same as ever but a little dirtier, her colors a bit more faded. It was my fault. I was neglecting her.

Sometime after my return, Grandmother gave me a white nightshirt that had belonged to Mother. Grandmother had brought it with her from Kabul, and now she told me she wanted me to have it.

NOT EVEN IN PAKISTAN were we safe from the long reach of the Taliban and their supporters. In April 1998, a year after my first mission to Kabul, a demonstration we staged in 102-degree heat in Peshawar turned violent, through no fault of ours.

Students from a *madrassa* Islamic school, their bearded faces twisted with hate, suddenly began pouring out of the school clutching sticks and batons and rushed across the street toward us. As we shouted our slogans for women's rights and against the bombings, the killings, the torture, and the rapes of women committed under the Taliban government in Kabul, the men in white turbans started lashing out at us.

I had never seen any demonstrations in Kabul, either under the Russians or since. I remember that Mother once tried to explain to me what a demonstration was, but I never quite understood what she meant. In classes, Soraya had told us about the many demonstrations in which she had taken part in Pakistan. "Strange, isn't it?" she said. "There I was, my face hidden so that no one could see my face and recognize me, but shouting slogans against all that the *burqa* represents."

For the protest in the town of Peshawar, several hundred RAWA members and supporters, both women and men, had been driven separately to locations close to the street we had chosen for the demonstration. It was in an area where many Afghans lived and worked. Some women had even traveled from Afghanistan to take part. It was only when we all reached the street that we unfurled the banners with the association symbol and our slogans.

"Down with fundamentalism!" the cry went up. "Women's rights are human rights!" "Long live democracy!"

It wasn't long before the cries of the protesters toward the front of the march suddenly took on an even more defiant note. I was walking with Saima, at the tail end of the throng of people. I knew something had gone wrong when a senior member came pushing through the crowd, calling to a male supporter.

"There's a problem at the front," she yelled at him. "Bring as many men as you can, and get them up here quickly!"

We were always prepared for trouble. Both RAWA members and male supporters had sticks ready in case we were attacked, and RAWA always supplied a few nurses.

I wanted to find out for myself what was happening. Saima and I struggled toward the front.

The students from the *madrassa*, set loose against us like dogs by their mullah, were kicking and whipping everyone in sight. Several of them were trying to seize the RAWA banner, but the women holding it were big and strong and would not let it go. There was a big tear in the cloth, and the women had cuts on their faces and arms.

Saima and I did what we could to help them. I took many blows and gave several. I saw that the arm of a friend of mine hung from her shoulder at a crazy angle. It was broken, but she continued to fight with her good arm.

I heard Zohreh, another member, call out to me. She was half lying, half sitting on the ground, her hand to her pregnant stomach. Her breath came out in short puffs. I saw that there was blood on her trousers, and thought that her legs had been injured.

We carried Zohreh to a nearby shop and called an ambulance, hoping it would be able to force its way through the crowd. We could not find the RAWA nurses and tried to give her some relief by fanning her with a newspaper and giving her water to drink.

Later, I found out that Zohreh had lost the baby. I learned that before the demonstration, other members had told her not to take part because of her pregnancy, but she had insisted that she must participate, it was important to her.

Not content with killing children and adults, the *madrassa* students who gave their name to the Taliban were even killing babies before they were born.

PART FIVE

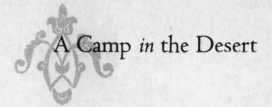

A Camp *in* the Desert

Chapter Fifteen

THREE YEARS AFTER I first joined RAWA, it sent me to live at a refugee camp that lies on the outskirts of the Pakistani city of Peshawar.

I arrived in a cloud of dust so heavy that even though I wrapped my scarf around my face, the dust got into my eyes and hair. To get to the camp from the city, I sat on the open back of a truck, riding next to the supplies with some other RAWA members. For the last half hour we bumped along a dirt road that stretched across the desert. Cars that passed us going the opposite direction blew up fresh clouds of dust.

I was greeted by Ameena, the RAWA worker at the camp. Three years older than I, she was tall and skinny, with long dark hair that she wore in braids. Ameena told me that she had first arrived at the camp as a refugee when her father brought her there as a young girl. He had sold vegetables at a market in Afghanistan until he lost his job. She had studied at the camp's school and since then has never left.

She showed me the house that we would be sharing. It was in the middle of the camp, a small dark mud house very much like the one I had been brought up—with bits of mud falling from the ceiling from time to time, just as they had done at home, and the same sound of termites munching their way through the table and chair. Piles of papers were everywhere. There was no heater. The toilet was a hole in the ground in a small shed outside. Apart from the papers, it was exactly the same kind of house that the refugees lived in.

We had only been in the house for a few minutes when the refugees started knocking at the door. The news of our arrival had spread quickly among them, thanks to the children who were always playing around the houses, and they wanted to welcome me with an embrace. It chastened me to realize that they believed we could make a real difference to their lives.

That evening, as we sat outside the house talking, I told Ameena that there were many more stars in the sky than I could see from the city. As a child, and with Grandmother's help, I had often tried to spot the star that shone brightest.

"Yes, the camp is so dark it makes the stars stand out. You can see everything," she said. "I often walk around the camp until after midnight. I'm always looking at the sky."

"But surely walking that late is dangerous for you. Someone could come up from behind and . . . "

Ameena nodded. "It's true that the camp is not always a safe place. We just have to be careful. But don't worry about me."

That night she wanted to give me her bed, but I refused and insisted she let me sleep on the carpet. I put my books and the little bottle of perfume I had brought with me on a small table. Although I knew I would never wear it in the refugee camp, I liked to see it there. From that first night I thought of the house as my home.

Early the next morning I saw much of the life of the camp pass outside my window—it was so busy it could have been a television set. Women, balancing milk from their goat on their heads, shouted angry warnings at the children who rushed past them, pulling and pushing at one another on their way to the school.

I was soon struck by how much compassion Ameena had in her heart—more than I would ever have. A few days after my arrival, we were due to hold an important meeting at the camp. RAWA members from other camps were coming for a big discussion on how best to run them. We were just about to start the meeting when I realized that Ameena was missing, so I went out to look for her.

The camp was huge. There were mud houses and tents for two thousand families, and they were set in a desert of dust, with not one green bush or tree in sight. I did not know where to look. It took me an hour to walk from one end of the camp to the other. I walked and walked, knocked on doors, asked dozens of people whether they had seen her, and sent children running to search for her. No one knew where she was.

I found her at last at the other end of the camp, her arms stretched around a dirty child—so dirty I would never have touched him—whom she had swung across her hips. He was in tears, his hair was matted with dirt, and the mucus from his nose was running down to his chin.

"Ameena, what on earth are you up to?" I asked angrily. "Have you forgotten the meeting? We're all waiting for you."

"Ah, I am sorry. I just saw this child. He's only seven years old, and already he's working in the brick factory. I thought I would take him to the store to buy him a sweet," she replied.

The factory just outside the camp was a terrible place. Every day it spewed black smoke over the camp as old tires were burned as fuel. A boy like the one Ameena was carrying would get up at four o'clock in the morning and work without a break until evening, first shaping the bricks into a muddy mess and then carrying them, still in their iron molds, to the oven, his back straining under the weight, his hands scratched and bloody—all for as little as ten rupees a day. Adults were paid sixty rupees a day, less than a dollar, providing they worked fast enough to produce five bricks.

Children and adults worked at the oven on even the hottest days of the year, their bodies baking along with the bricks. They were treated like slaves. The work halved their life expectancy because of the dust and smoke they breathed in from the oven. The Pakistani owners of the factory never allowed us to visit it.

I could not be angry with Ameena, and we laughed as we made our way to the meeting with me doing my best to hurry her up. I always call Ameena "turtle" because she walks so slowly, almost as slowly as Grandmother.

Often I would find Ameena sitting with the women, crying as she listened to their stories. She was always willing to listen to them and became so wrapped up in their stories that she would forget the work she was supposed to be doing that day.

I HELD THE HANDS of the old woman, but they were so cold it was like clutching a corpse. She gave no sign that she was aware of my presence. She did not cry, she simply stared at the floor. Her head, wrapped in the black scarf of mourning, hung down to one side. Her face was white, her lips purple and caked in blood. The only movement was her thumb rubbing slowly against her index finger.

The men who had arrived with her at the refugee camp had told me that she had lost her son Najib. I knew Najib. He had lived and studied in the camp before becoming a supporter of RAWA and starting to teach classes to the boys. One year earlier he had left to take care of his mother in Yakaolang, a town in the center of Afghanistan that is populated by the Hazara tribe, which the Taliban hate.

A few days earlier, I had heard about the massacre there on the radio. I caught only part of the news—the radio is very old. It usually takes at least half an hour to find the right wavelength, and even then, the signal comes and goes. It never fails to make me furious.

When I first saw the refugees arriving, I was at a loss for words. There were three dozen or so, mostly women and children. We tried to do what we could for them, but the adults all sat as if in a daze, resting their chins in their hands. They had lost everything. They were in a world of their own.

I did not even try to say the words one usually says to someone who has lost a close relative or friend—"I am sorry, and I hope that you will be strong enough to survive this tragedy." It would mean nothing to them, given their suffering, and they probably wouldn't even hear me.

The mother, a scrawny, disheveled figure, was sitting motionless in a corner of the room. Next to her was the young woman who had become Najib's wife only two weeks before his death. Her hair and her hands still bore traces of the rust-brown henna she had worn for her marriage. I found it impossible to look at his mother because my mind was so full of Najib, remembering how hard he had studied and how he was always ready to help.

I was too overwhelmed to stay in the room, so I walked out without saying anything. I knew that at that moment I did not have the strength to help them. Sometimes I think that even if I sat for days on end listening to the terrible stories the refugees have to tell until my ears dropped off, I might become crazy myself, but I would not change anything in their lives. So it is better to get on with my work. Better to show that we are working to help them, rather than promise them the world.

Eventually I learned the details of the massacre, which took place in January 2001, from a RAWA friend who was collecting the survivors' accounts for a report. I did not take part in this. I could not bear to sit and listen to the men who had survived describe the killings, imitating as they did so the sound of the Kalashnikovs with their tongues, while the women around them who had lost sons, husbands, and brothers relived their agony. It was the sound that they had been forced to listen to as they cowered in their houses. At each volley of bullets, they pictured in their minds their men falling to their knees.

For me the job of interviewing refugees to write reports for the publications is one of the worst jobs in RAWA. I can't do it. Although I'm sure no one can tell by looking at me, because I don't show my real feelings, there is so much happening inside me when I have to hear one of these stories that I am afraid of how I will react.

A Taliban commander had retaken the town and ordered that all boys and men between the ages of thirteen and seventy who were considered to be anti-Taliban be rounded up. The commander's soldiers herded the men to assembly points, put them before firing squads, and executed them. About three hundred men were slaughtered. One teenager was skinned alive. The killings were apparently intended to deter people from cooperating with the Taliban's enemies in the future.

One man who had escaped told me that after the killings, he had taken Najib's mother by the hand and brought her out of Afghanistan. "She never said a word to me. Not even to ask where I was taking her," he said.

Three days went by before I could bring myself to visit the mother. I sat down beside her and felt her pulse. There was almost nothing. Her daughter-in-law, who was seventeen, sat in a corner, glancing occasionally at me.

"Mother," I began, "I have no words to express our grief. Najib was a brother to us. He helped us, and as long as we are alive, you can count on us to help you. Even to seek revenge for the blood of your son."

At that moment I would have done anything for this woman, even pick up a gun. I wasn't thinking straight, and I should never have said those words. Sometimes my feelings get the better of me and I cannot help speaking out. The mother was still immobile. There was no life in her eyes. I could not tell whether she welcomed my words.

"Mother, this is your home now. We will look after you. Najib is not alive, but his friends will always come to see you," I continued. I talked and talked, but she never acknowledged me.

Eventually I stood up. "I am sorry to have spoken so long," I said. "I know this is no time for you to hear me."

There was still no indication that she had. Nor did she move as I walked out of the room. I had taken only a few steps down the corridor when I heard the scream, a shrill scream that hit me like a slap in the face.

"Najib!" the mother cried. "You have killed me, you have killed your mother!"

I ran back into the room. The mother who had been so still only a few moments ago was in a frenzy, beating herself and pulling at her hair as her daughter-in-law struggled to calm her down. I sent someone to fetch a doctor, then took hold of the mother's hands and managed to get her to sit down. Her pulse was so fast I thought she would die.

Suddenly she stopped shaking, as if her energy had all been spent. She placed her hands on my cheeks and kissed the top of my head. In return I kissed the back of her hands.

The next day I learned that she was refusing all food and drink. I went to see her to try to make her drink some milk. As her daughter-in-law held her head steady, I brought the glass to her lips. The only result was a thin film of white over the crusts of dried blood on her lips. The doctors put her on intravenous feeding.

I never spoke to her again about her son's death. But I could not get her out of my mind, no matter how I busied myself with other work in the camp. An old nightmare returned. Someone or something is coming toward me. I know it wants to hurt me. I am powerless to shout or move. I open my mouth but can't make a sound. My legs won't move. Something is closing in on me, but I have no idea what it is. All I know is that it is black. It comes closer and closer.

I wake up, and because I know that I will not be able to sleep again, I switch on the light and go and do some work, whatever the time. Usually it is two or three o'clock in the morning. I worry that one night the something in my dream will get so close that it touches me.

When I spoke to Najib's widow, who I knew could neither

read nor write, I told her, "Don't wear the black clothes of mourning. If you want to show your love for Najib, why don't you do something he would have been proud of—why don't you join the literacy classes at the camp? He would be happy to see you do that."

"I will try," she replied.

She started the next day. It was the only time during the day that she did not spend at her mother-in-law's side. She was a good student, although she worried constantly about her marks. We had given them a room in the orphanage, the only one we had available. It was so hot that they had to leave their door open, with a curtain hung across it in an attempt to keep out the flies and mosquitoes.

We asked all the girls in the orphanage to take special care of the mother, and not to burst in on her by suddenly pulling aside the curtain. We asked the girls to try to persuade her to drink some milk with them every day. Little by little, she started eating again.

Chapter Sixteen

WE THOUGHT we had organized everything properly. We had distributed coupons to the most needy families in the camp, saying clearly when and where they should come to collect one of the fifteen hundred blankets we had to distribute, each piece of paper valid for one blanket only. We had negotiated with Pakistani officials for permission to use a small building in the camp, which belonged to them. We particularly wanted it because it was set in a compound surrounded by a wall, which would enable us to control the flow of refugees coming to collect the blankets. Or so we thought.

Hours before the distribution was due to start, several hundred refugees had gathered outside the gate to the compound. Ameena and I, together with a dozen other RAWA members, spread piles of blankets around the main room of the building and told our male helpers, who were standing by in case there was any trouble, to let the first refugees in.

We had barely handed out the first dozen when we heard shouts coming from outside. "Why are you giving the blankets to some people and not to others?" I heard a man cry. "We all need blankets!" someone else shouted.

Within seconds, the men we had stationed at the gate were overwhelmed and the crowd charged toward the house. Pushing and shoving one another, men in rags and women wearing veils or burqas fought their way through the door and into the room. The handful of male supporters who were in the room with us tried to push them back, wielding their batons to try to form a barrier in front of us. We urged them not to hit the refugees hard, but they told us it was the only way to protect us.

These people were driven wild by desperation. I was pushed back against the wall, and so were the other girls. A few steps in front of me, an old white-haired woman thrust out her piece of paper. When I shook my head, unable to move and get her a blanket, she threw herself to the floor, screaming her anger. Some men tried to calm her down.

Nearby, a man and a woman pulled on a blanket as they struggled to push each other to the ground. "My children are dying of cold!" the woman screeched. The blanket ripped in two, and they started fighting over the pieces. Soon, bits of cloth were drifting about the room above the heads of the men and women who were now destroying all the blankets it contained—so many bits of cloth and fluff that it became difficult to make out the refugees. It was like the sandstorms that sometimes blew across the camp making all the houses and tents invisible and finding its way into our house to smother everything in grains of dirt and dust.

We had no choice but to flee. I felt no anger against the refugees. I couldn't help thinking that if I had been in their place, I would have done the same.

Later, I saw the white-haired woman striding across the camp, proudly carrying a torn piece of blanket on her head as if it were the most valuable thing on earth.

That evening, Ameena looked exhausted. She hardly touched our dinner of rice and vegetables. Her face was very white, and her eyes looked strange. I said something to her, but I could see that she was not following. When she got up to take our plates into the kitchen, they suddenly slipped from her grasp and she slumped to the floor. As I stared, her body started shaking. Her head rocking backward and forward, she pulled at her hair. I could see strands of it in her hands.

I knew that it was an epileptic fit and that I should force something into her mouth because she risked biting her lips or suffocating on her tongue. I grabbed a spoon, but there was so much power in her that she was too much for me and I had to scream for help.

When she became calm again, she turned to me and smiled. I was so tense I tried to joke about what had happened. "You just made me go completely crazy, and now there you are smiling," I said.

She sat up. "I hope I didn't hit you or bite you."

"I think you were doing it on purpose. You just felt like getting at me," I said, smiling.

Before falling asleep, she told me about her family. She was no longer on speaking terms with her father, who had returned to live in Afghanistan two years earlier. "He wrote to me to say that my mother is sick, that my brothers are also angry with me, and that I should go back and marry. My father says I have to respect his authority. But how can I leave here? I feel desperate if I miss even one day at the camp," she said.

Ameena always apologized to me after her fits, and we always joked about them. Even for Ameena, whom I see as the strongest person in the world, the weight of the lives of the refugees sometimes becomes too heavy a burden. I tried to persuade her to see a doctor, but she always told me that she felt fine, and that there was no need. Whenever I left the camp I tried to bring back some fruit for her. Once, when she was alone in a room at a house in the city, she fainted, her hand resting on a heater. When a friend found her, Ameena's hand had severe scorch marks on the skin.

I never suffered physically as much as Ameena did, but sometimes my heart would start beating very, very fast. I would not be able to move, I would feel a pressure on my chest as if someone were pressing down on it with all their strength, and I would start sweating. Then, just as suddenly, the rhythm would change until the heartbeats came slowly, much more slowly than normal. I would just wait for it to pass.

One of the most common diseases at the camp is malaria, which is spread by the mosquitoes that plague us day and night. I've caught it three times. The worst time was when I was about to attend a meeting in the town and my hands started to tremble. I had felt sick in my bones earlier that day but had promised to go to the meeting all the same. Still I could not stop my hands from trembling. I felt cold and hot at the same time.

My friends realized what was happening and covered me with a mountain of blankets even though it was early summer. I was in a high fever and started vomiting. My friends took me to the hospital, where I was told that I had caught malaria. I lost five kilos but recovered quickly. Malaria is normal at the camp, and our doctors always have malaria patients to care for.

ONE SUMMER MORNING I watched as Fatima, an Afghan doctor in her mid-thirties, prepared to receive patients in 104-degree heat. Fatima's office was an old chair set under the beating sun in the dirt between two rows of mud houses. Her dispensary was the open trunk of a car where a pharmacist waited to hand out the medicine she would recommend. Her uniform was a scarf that made her all the hotter but that she had to wear around her head so as not to offend the refugees.

The lack of facilities did not matter to the old women who had scuttled like ants from all over the camp to wait in an untidy huddle for the doctor. For them the doctor was a goddess with the power to change their lives.

An old woman, her face as brown and wrinkled as a walnut, was the first to get to Fatima, and she started speaking without any form of greeting. "My daughter, please help me, I am sick," the old woman said.

Fatima did not bother to ask her age, as the old women usually have no idea how old they are. But before she could ask what the woman was suffering from, the woman continued, "My daughter, I am weak. Give me the medicine."

Many of these elderly refugees believe there is only one kind of medicine that cures all ills. It took Fatima several long minutes to find out what the woman was suffering from, and the line of people around her was swelling all the time.

"Everything is wrong with me," the woman kept answering. "All the sickness of the world is in me, in my bones, and I am weak. Give me the medicine to make me strong."

Finally Fatima found out that the woman was deficient in proteins and vitamins because of her diet, and gave her a slip of paper to show the pharmacist. For longer even than it had taken to establish what was wrong with her, the old woman seized Fatima's head in her bony hands and kissed it. She thanked Fatima again and again, wished long life to her and her children, and was still praising Allah for his mercy when one of the camp workers gently led her away.

It took the pharmacist all his powers of persuasion to convince the old woman that no, she should not take all her medicine in one go and that if she did it could kill her. When he told her to take the pills three times a day with her meals, the old woman looked blank—she did not have three meals a day. She counted herself lucky if she managed to have one. So the pharmacist timed the doses by the calls to prayer from the small mosque in the camp. "Mother, take the first pill after the second prayer . . . ," he told her.

Fatima was already listening to another refugee, and there were hundreds more waiting. Fatima knew that she would have to listen to them tell her not only about their ailments but also about their lives. "Doctor," they would say, "I have lost everything. My brother has disappeared in the fighting." And Fatima would listen, because it is she who has chosen to abandon her white blouse and a quiet, clean visiting room in the city to spend her days among the poor and the dirty who worship her like a deity.

I watched Fatima, and I thought of the clandestine medical teams that RAWA would send to the more remote areas of Afghanistan where, under the Taliban, women were dying of curable diseases simply because there were no women doctors to see them. The teams drove into the most desolate villages, using ordinary cars because an ambulance is too easy to identify, and tried to spread the word that they had arrived.

The villagers were so happy at their visit that when a doctor entered their home, a child would bring a basin of water with a towel so that the doctor could wash her hands. It was a long-established tradition in the most rural villages, but it had little to do with hygiene, as no soap was used. I disliked it because it was a legacy from feudal times, a sign of submission.

Often the medical team would sleep in a tent because they did not want to expose to danger someone whose hospitality they would otherwise have accepted.

I hope that one day I will be able to study to become a doctor myself.

I HAD NEVER thought that I would play the part of jailer, but it was the only way to ensure that the women would keep attending the literacy classes that Ameena and I had started up in the camp. So I always slipped the heavy chain across the door of the classroom and locked it from the outside because without the lock, children would be slipping into the class every few minutes to stand in a corner, pointing and poking fun at the sight of their mothers perched on the small chairs designed for kids.

For weeks, Ameena and I had spent our evenings walking up and down the camp trying to find enough women willing to come to the classes. We always tried to speak to the women on their own because we had to persuade them before they mentioned the plan to their husbands, as they felt they had to do. Most of the women simply laughed in our faces.

At seven o'clock one morning, a woman knocked at our door and brought us a breakfast gift of one egg. When I held it in my hands, it was still warm from the hen. I offered it to Ameena, my hands stretched out, but she pushed them back toward me. It was a particularly welcome present because breakfast usually meant just one or two slices of bread—we had no butter or jam—and some tea.

We showed the woman in and offered her tea, and she sat herself down. But it was some time before she explained why she had come to see us.

"My daughter is going to the school, and I have seen how fast she writes. I am thinking of going to the school too. But I am worried that people will wonder what a woman of my age—I am fifty-five years old—is doing in a school when she should be at home and praying. My hair is gray. Is it shameful if I start at this age?" the woman asked.

I clapped my hands and assured her that she should feel no shame, only pride.

"But my daughter and the other children will laugh at me," she said.

"You must give it a try. What about the shame you feel when you see your daughter write and you know that you cannot even write your own name? Imagine, you will be able to write letters to your relatives and your friends in Afghanistan, and someone will read out your news to them," I said.

I offered to accompany her to the class, which we were starting that afternoon. She turned up as promised, although she told me that her husband had poked fun at her, saying that he did not know he had married a great writer and philosopher.

We had managed to get together a little group of courageous pioneers, and just before four o'clock, one by one they slipped into the classroom where their children studied in the mornings, holding their veils over their mouths, as embarrassed as if they had been caught stealing their neighbor's chicken.

Several of the women had come without telling their children, but the word had got around, and shortly after Ameena started distributing a notebook, a pencil, a sharpener, and a ruler to each of them, some of the women's daughters turned up, darted into the room, and started giggling. Ameena asked me to chase them out and lock the door so that she could continue the class.

It was some time before the mothers had the courage to ask their children for help with their homework. Some of the mothers, when they were finally able to write their names, thought that the lessons were over and that they now knew everything they needed to know. We managed to keep them coming, partly by giving them rewards—some soap, a kilo of rice—if they did not play truant.

The literacy classes were important to us because they were not just about learning to read and write. We used them to teach the women about the different kinds of contraception available to them and tried to discourage them from having many children. The shame that so many women felt about the literacy classes never totally vanished, but gradually we found more and more candidates. Even some grandmothers came to sit on the little chairs. I was proud of them. I felt that they were braver than I would have been under the same circumstances.

SOMETIMES IT WAS HARDER to get the children to school than their mothers. One little girl kept coming to me in tears, saying that her father—a man who had served as a soldier with the Mujahideen—was beating her because he did not want her to go to school.

After asking her mother for permission, I went to find the father. He wore a white turban, and his manner was strange from the moment he opened the door. I was wearing a small scarf over my hair, but he did not even look at my face. He simply stepped back and announced to his wife, "There is a woman at the door." His wife, who was wearing a big scarf over her head, led me into the tiny room they shared with their six children. There was a cloth embroidered with a prayer to Allah on the wall.

I explained that I had come to talk to him about his eldest daughter. "What do you want?" he asked sharply, still avoiding my eyes and without offering me anything as would have been the custom.

"Your daughter is crying every day because you do not want her to go to school. But you must realize that if you love your daughter, you should allow her to study. If you don't allow her or any of your children to study, it will be bad not only for their future but for yours too," I said.

I might as well have been talking to the dry mud wall behind him. "But I am a Moslem," he said, "and I want my children to be good Moslems too. Why do you not teach only the Koran? Why do you teach these other subjects too? You are infidels, and I do not want my daughter to become an infidel."

I thought of the *madrassa* schools that Grandmother had told me about during my childhood in Kabul and more recently in Peshawar, which taught only the Koran and from which the Taliban had emerged. "The families who are sending their children to the school are also good Moslems. Good Moslems are allowed to study other things as well as the Koran," I answered.

I was becoming increasingly irritated, not only by his words but also by the fact that he kept staring at the ground as if I were too distasteful to look at. He knew very well that we respected Moslem culture. When we showed videos of Western films to the children, for example, we always censored them first. One of the best films we got for the camp was *Schindler's List*, and we cut a scene that showed a man and a woman in bed together. If we had left it in, the children would have talked about it to their parents, and the parents would have wondered what kinds of films we were showing.

But my arguments that knowledge was power, that knowledge could give his daughter a chance of a better future, had no effect on him.

"She is my daughter, and I will decide whether her future should be dark or bright. She is of more use to me using her little fingers to weave carpets. Now you can leave," he snapped.

I got angry. "I came here for the sake of your daughter. You do not know what is best for her future. She wants to learn. She is lucky to have a school here. But you want her to be ignorant. If you beat her again because you do not want her to study, we will take steps to have you removed from the camp."

I was furious with him, and furious that his daughter should pay such a heavy price for his beliefs. She stopped coming to the school entirely.

ONE OF MY DREAMS is that every town and village in Afghanistan should have access to a library with many, many books—books on all the sciences, literature, and art in both the Persian and Pushto languages, books that will document the heritage of Afghanistan and what the Mujahideen and the Taliban did to it.

It was in the camp that I met a woman from the province of Bamiyan who had lived within sight of the two giant stone statues of the Buddha that the Taliban shot to pieces with their artillery. "I woke up every morning with the mullah's cry," she told me, "and every morning the first things my eyes went to were the Buddhas. They had gone, after one and a half thousand years. Now they will never come back."

Chapter Seventeen

"AUNTIE, I need a new pair of shoes."

Shamms stood at my door, the child I love the most in the whole camp. Shamms is a beautiful, very little five-year-old of the Hazara tribe, the most persecuted tribe in Afghanistan. He has the blond hair and Chinese-like eyes typical of that ethnic group. He is the only boy in the girls' orphanage. He is too young to be with the other boys, who are much older than he is. No one has yet told him why he is in an orphanage.

No one has told him that his parents are dead, that they were killed when a bomb flattened their home. He thinks the fourteen girls in the orphanage are all his sisters, but in fact, only one of them is his real sister, and I see him spend more time with her than with the other girls.

After the parents died, Shamms's eighteen-year-old brother brought him and his sister, who is one year older than Shamms, to the refugee camp. The older brother told us that he had nothing, that he could not take care of the two children. Shamms has no doubt that the RAWA member with whom he lives is his mother.

"Auntie, I need some shoes," he repeated. "I need them for soccer." He always calls me "Auntie."

There is nothing Shamms loves more than shoes. I checked with his teacher at the orphanage, and she told me that he had several pairs. "Don't give him any more shoes. He just doesn't need them," she said.

I knew it was a waste, but sometimes I cannot refuse him. He was convinced that with a new pair of shoes, he would be able to score more goals and one day play for the Kabul team— if there ever is a Kabul team—or even for Afghanistan.

"All right," I said. "I will get you some new shoes. Is there anything else you need?"

Shamms looked shy and shook his head. I always treat Shamms as if he is my superior. I even offer him tea when he knocks at my door, something that is usually only offered to adults. Shamms always refuses.

Because I treat him with the respect due to an adult, Shamms thinks I cannot be very important in the camp. *Surely*, I can see him thinking, *the important people are the ones who are more strict with me.* When he comes to the house, he usually starts by asking for Ameena, because she really is an important person.

Shamms is bright and cheerful. He has just started learning the alphabet, and he rushes to me at all times of the day saying, "Auntie, Auntie, can I read to you?" He stands on the tips of his toes to kiss me, and I tell him, "One day, Shamms, you will no longer want to kiss me." He has also started learning English, and he is never shy about getting up in class when a visitor is shown in and announcing, as he tries to stand as tall as he can, "My name is Shamms."

Whenever I leave the camp, I try to bring him back some stationery because he loves drawing. And he loves books, books that he cannot read yet. He can spend a whole hour playing with a chemistry book, turning the pages backward and forward although he cannot understand a word of it.

Whenever he can get hold of one rupee, he runs to a shop and buys four tiny firecrackers. They make a very bad noise, especially for the refugees who are living in the camp. They lived through blasts and explosions in the country they have fled, and now they have to endure these firecrackers. Often they come to me to complain, but like them I have to endure this noise even when I am trying to get some sleep.

Once, I was showing a Western journalist around the camp, and a firecracker went off at his feet. The journalist jumped with fright. He thought he had been shot. I saw Shamms standing a little way away with some other children. I was furious with him and shouted at him, "Stop that, you donkey!" He is crafty. He is very careful to never let me catch him with the firecrackers in his hands. But I knew it was him.

Shamms loves to make a noise. He loves the RAWA songs and listens to them with the volume turned as loud as possible on the cassette player. He is too small to pronounce the words properly, let alone remember them, but whenever I ask him to sing something he will just go ahead anyway and invent new words for the song. Still, when we have to give a newly arrived refugee, like Najib's mother, a room in the orphanage, Shamms quickly senses that he must not disturb the guest and he makes as little noise as he can.

Sometimes I give some money to Shamms to buy not fire-crackers but a kite. He tries to fly it on his own in the middle of the camp, but he is too small and he is not yet able to make it take off or stay in the air. After a while, there is a big tear across the thin paper.

I don't believe in giving up a little boy like Shamms for adoption. We get many requests from couples in America and Europe who would like to adopt one of our orphans and offer them a different life. But we only accept long-distance adoption, with adoptive parents helping to bring up a child from afar. These children are the future of Afghanistan. They have many talents, and if we send them to the West, their talents will be swallowed up there. If we concentrate so many efforts on their education, it is because we want our children to build Afghanistan's future. We cannot afford to lose a generation.

Even if Shamms were to be offered simply the opportunity to go and study in the West—a chance that I was offered when my cousin came from Canada to see me at school—I would advise him not to go. If a child like Shamms were to spend even a few years in the West and lose touch with his origins, he would never want to come back to the dust of Afghanistan.

Some people tell me that Western couples can give these children the love of parents. But RAWA provides this. Our members are like mothers to these children, and the male supporters are like fathers. Many of the children are taken in by the families in the camp and grow up as if they had always been part of the family.

AMEENA AND I were poring over the end-of-month accounts of the camp. It was a slow business. I was still as bad at math as I had been at school, and the electricity had been cut off again that evening so we had to work by candlelight. Then there was a light knock at the door of the house.

I went to open the door and saw a little girl standing there. "Come quickly," she said. "There is a woman who has tried to make herself die."

When we reached the woman's house, we discovered her lying on the floor, her head resting in a small pool of blood and vomit. In a corner, I saw a bowl of water in which some bits of bread, which had perhaps been thrown away by another family, were floating so as to make them soft.

The woman was covered in sweat although it was a cool day, and the skin around one of her eyes was battered and bruised.

We found out that she had swallowed half the contents of a bottle of rat poison. An ambulance took her away.

It was only when we visited her in the hospital, where she was recovering after the doctors had emptied her stomach, that we learned why she had tried to commit suicide. Her husband, Mohammed, was an opium addict, and although they had no money, he refused to let her out of the tent where they lived to try to earn something or even beg for alms. When she asked him for money to pay for some food, he would beat her.

I went back to the house to find Mohammed. He was from the Kunar province in the north of Afghanistan. Slowly, very slowly, he told me that he had discovered drugs at the Karkhano market in Peshawar, a no-go area for the police where addicts huddle together and use heroin.

"I went there after working for a month in the brick factory," Mohammed told me. "The pain in my back was so great that I could not lift even one brick. I could not work there anymore. I met a dealer, and he said that I should forget about looking for a job. I remember he said to me, 'You will never find a job in Pakistan, and you will never be able to return to Afghanistan. But I will give you something that will make you forget all the problems you have in this world because of money and work and your wife.' "

Mohammed paused, remembering the first time. "I went into another world, a beautiful world. I was so happy. I was not thinking about poverty, about what my wife and I would eat. I was happy in my heart. For maybe two or three hours I was happy, then I needed some more. When I got back here, my wife said there was nothing to eat. I had nothing in my pocket. So I beat her."

He returned to the Karkhano market only a few times, because soon he could no longer afford the heroin. He switched to opium. Now he had no job, there was no food in the house, and he was addicted.

I asked him why he did not allow his wife out of the house.

"Because it would be shameful for me," he replied. "How can I let my wife go out to find money to pay for the opium, when I have not got the strength to leave this house?"

When I told Ameena what Mohammed had said, she was particularly saddened. Her brother, who was still in Afghanistan, had also become an addict because he had no hope of schooling or of a future. Both men were among thousands exploited by the Afghan and Pakistani drug barons, who plied their trade with the help—at a price—of the Taliban.

We made a collection among RAWA members and got enough money to pay for Mohammed to be treated for his addiction. Then we persuaded him to return to the brick factory. Although it was painful work, it was the only job he could get.

FEW THINGS at the camp anger me as much as the wedding ceremonies I see some of the poorest families hold. One of the most shocking for me was the wedding of a teenage girl who was being taught by Ameena. The girl's mother attended

Ameena's literacy class, so there was no way that Ameena or I could refuse the invitation.

I was not keen to go, partly because the girl was being given in an arranged marriage to a man more than thirty years older than she was. The family was so poor that in the days before the wedding they borrowed some carpets and other things from us in preparation. When we arrived at the wedding, we found that the parents had invited all their relatives, not only from the camp but also from the city, and because there wasn't enough room in their small mud house, they borrowed some space from their neighbors.

The air was heavy with the smell of the goat and the two sheep that they had slaughtered to feed their guests. The food had cost the family a great deal of money. A blind old man had been hired to sing traditional songs while someone played the *dootar*, a two-string guitar. Because he could not see, he was free to sing both in the rooms set aside for the men and in the separate section where the women had to stay. In that kind of traditional environment, a seeing man would not be permitted to watch the women dancing.

When I went to congratulate the bride, I saw that she wore a red dress laced with gold, heavy makeup, and was covered in rings and jewelry. Ameena said to her, "Promise me that you'll keep coming to the school."

The girl looked abashed. "I don't know. It depends on what my husband says," she replied.

"No it doesn't," Ameena retorted. "It depends only on you, and it's your future that is at stake."

The guests were so poor that they gave the wedding couple presents of ten or twenty rupees—just enough to buy some eggs and some milk. The celebrations lasted for three days.

The following week, we went to see the mother. The family had spent so much on the wedding that she did not even have any sugar to put in the tea that she offered us.

I was upset and disappointed in her. "Mother, first you decide to give your young daughter to an old man, and then you spend so much on the wedding that you can't afford sugar in your tea?" I asked her.

I saw tears in her eyes. "I had no choice. We had to find someone to protect my daughter. My husband had to borrow so much for the wedding, because we could not have lived with the shame of something small," she said.

I thought of my parents' wedding and Mother's insistence that it should be exactly that, "something small." I wondered at how an Afghan family could hold such a celebration when there were so many bombs exploding every day in their country.

The daughter did not return to school. When we went back to the house to see her, she told us that she was pregnant and that she would have to wait for the child to grow up before she could go back to class. I hope that one day she will come back.

I HAVE ALWAYS avoided seeing people die. I think I can look at pretty much anything, but not someone dying, and not a dead body. Still, sometimes I have no choice, I have to pay my respects. The refugees would quickly lose their regard for us if we attended only their weddings and not their funerals. It would be like sharing only their moments of joy and pretending that the sad moments did not exist. But it is an ordeal for me every time.

When I am in the room where the body has been laid out, I look everywhere—at the mourners, at the floor, at the walls—everywhere except at the body. Once, I saw a woman who had been seriously wounded in the stomach by shrapnel. Her husband abandoned everything in Kabul to take her to a hospital in Pakistan. He was shouting while the doctors tried to save her, and I could do nothing to help him. I saw her dead body afterward, but I could not bring myself to help wash her for the funeral.

Once Ameena and I were called to a tent in another camp that we were visiting. We found a man was stretched out on the ground. He was very sick and had tied a string around his head, the way it is tied on a corpse to hold the mouth shut. I took the string off, and he told me, "I am sick and I am dying, but I have no one to tie this piece of string on me. So instead of dying like a dog, I did it myself." When he died a few days later, I could not bring myself to go and see him. I can't even stand seeing dead bodies on television.

In the evenings, if I have the time, I sometimes go for a walk around the camp. I visit families and see whether they need anything, because often the families are too proud or too ashamed to come and ask us for something. When we know someone is sick, it is up to us to go and ask the relatives how he or she is. Once, I wore a *burqa*, because I wanted to see how the refugees were living without their recognizing me as a member of RAWA.

I always prefer to turn up without notice, because otherwise the family will prepare a meal for me out of their great sense of hospitality, and I do not want to take the food from their mouths. I am still angry with myself for once being so busy seeing refugees that I forgot that a woman in the camp had invited me to her house. She was poor, but she had prepared expensive food—chicken with rice—for me. I kissed her hand to apologize, and she forgave me.

My route often takes me past the cemetery of the camp. There are thousands of graves there, set in the dust, and the cemetery seems to have got a little bigger every time I walk past it. Some of the dead are from families that are so poor they have no money to pay for a piece of wood to record their loved one's name. I never go inside the cemetery. But I know that if we stopped doing our work, it would be twice as big as it is now.

When, in the small hours of the night, I fall asleep, I always worry about the safety of my RAWA friends, both in Afghanistan and Pakistan. I know that wherever they are, they can never be safe. My worries increase every time I hear a frightening story from the refugees or from RAWA members. One member told me after returning from Afghanistan that she had seen a woman climbing across the rubble of what had once been her house after it was hit by a bomb. Her son's head was on one side of the rubble, and the rest of him was on the other side. And this woman was trying to pick up the pieces. I saw this woman again and again in my nightmares. I cannot move or shout as I watch her.

I do not believe, as the Taliban do, that death can be a blessing, nor do I believe in martyrdom. I am frightened of dying, but only for one reason: I am afraid that I could die without helping my people or leaving some kind of mark. I would hate to die in a car accident tomorrow. When I see the ocean of pain and sorrow that my country suffers, I feel that what we are doing is smaller, much smaller, than a drop of water. But I must continue my work because I believe in it. I believe that it does make a difference.

Sometimes, simply because we cannot afford the cost, I am forced to refuse an appeal for help. I will never forget an old woman who came to me after fleeing her village in the center of Afghanistan because of the fighting. She begged me to allow her and her five children to be admitted into the camp.

She had lost her husband, and she bent down to the floor to touch my feet as she pleaded with me. "Please take us, or at least please take my children into your orphanage. I cannot feed them. They are yours," she said.

I had to refuse her. We could not afford to admit anyone else into the camp. We did not have the money to pay every day for their food, for their medicine, and for the children's education.

I could have taken her and her children into the house where I was staying, and I did once take a family in, but this would have been only a temporary solution. The woman was not angry with me, but I felt responsible for what would happen to her and her family. I do not know where she went. I hope she managed to get into another refugee camp.

PART SIX

Beyond the Veil

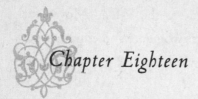

Chapter Eighteen

I WALKED IN through the small door of the Taliban embassy in Islamabad, found the women's section, and joined the line for passports. It was then that I noticed the sign hanging from the wall behind the employee at the counter. It was handwritten in black ink and beautiful calligraphy: *A woman in a* burqa *is like a pearl in an oyster*, it read.

I remembered how I had felt wearing a *burqa* in Kabul, and wondered how anyone could find such poetry in it. To me, a woman in a *burqa* is more like a live body locked in a coffin. But at least the Taliban could not stop Afghan women from traveling on their own outside the country.

I PACKED both my passport and my *burqa* when I made my first visit to America.

Eve Ensler, the playwright who wrote *The Vagina Monologues*, had invited RAWA to speak at a meeting in Madison Square Garden in February 2001, hosted by the V-Day Movement, which fights to end violence against women. Women representing associations from all over the world would get a chance to speak.

I had seen New York before in films on television in Pakistan, but when I got there I walked around in a daze. In my country, everything had been destroyed. I wondered how long it would take before even one building that could vaguely resemble what I saw before me would be built in Kabul. I thought of the energy and the work needed to bring this about, and realized that it would take centuries, and I would never live to see it.

The skyscrapers seemed like mountains to me. I dreamed that one day we would build a new Statue of Liberty in my country and that it would have the same meaning to Afghans as it did to Americans.

I noticed the wealth in the shops. I would have liked to buy medicine for the refugees, and small cameras that our members in Afghanistan could use. I noticed the happiness of the people, so busy at their jobs, free to worry about what they should cook for their guests that evening, and of the children, who could look forward to years of school and then college.

I met Eve Ensler and stayed at her home in the days before the meeting. She cried and hugged me when she saw me. Eve is very brave and a strong supporter of our cause. I first met her in Pakistan before she crossed the border with a RAWA member to talk to women in Afghanistan and to see our clandestine classes for children. She wrote a poem afterward called "Under the Burqa." She told me that she had asked me to bring mine on my visit because she wanted to use it for my speech.

I met Jane Fonda before the gathering and told her how much I and the other girls at school had enjoyed the film *Julia* and that it had inspired my work. I asked her whether she would like to make a film about Afghanistan, and she said we should discuss it. She was very kind to me. When I told her about what was happening in my country, she wept.

When the time came for me to go onstage, after Oprah Winfrey had read "Under the Burqa," all the lights went off save for one that was aimed directly at me. I had been asked to wear my *burqa*, and the light streamed in through the mesh in front of my face and brought tears to my eyes. A group of singers was singing an American chant, a melody full of grief, and I was to walk as slowly as possible—one step and then pause; and again one step and then pause. I had to climb some steps, but because of the *burqa* and the tears in my eyes, which wet the fabric and made it cling to my skin, I had to be helped up the stairs.

Slowly, very slowly, Oprah lifted the *burqa* off me and let it fall to the stage. It was the first time I had spoken in front of a crowd of eighteen thousand people, but I wasn't nervous. In any case, it was so dark out there that I couldn't see them. When I finished speaking, the lights came back on and I saw that the people had got up from their seats to stand while they clapped. I was happy that they showed their solidarity in this way, but for me it was more important that they would be inspired to help.

SEPTEMBER 11, 2001. I arrived at the airport in Islamabad, accompanied by my RAWA friend Saima and a male supporter who was our bodyguard and driver. I was due to fly to Spain to attend several conferences at which I would represent the association.

When we entered the main hall of the airport, we noticed that many people were crowded around a television set that hung from the ceiling. None of them were speaking. I got closer and read the headline: "CNN breaking news—Attack on America." There was nothing else on the screen, just dust, a lot of dust.

Soon the dust was replaced by the film of the twin towers.

So many times I had seen violence and terror in my country, but it had never been shown on television like this. I saw a man jumping from one of the towers. It was my nightmare. I imagined being in the tower, the tower that to me was like a mountain. I heard on the news that a man had called his mother, and I thought of my parents. I imagined that the planes hitting the towers were like bombs hitting the shelter under my house in Kabul, or the thousands of other shelters where people thought they were safe.

The commentator mentioned that the attacks might be linked to Osama bin Laden.

I went to phone one of the RAWA safe houses. I asked them if they had seen CNN or the BBC. They told me the television was switched off. I told them to switch it on immediately and that I would call back later. I no longer knew whether I should go on my trip or not.

We were all convinced that bin Laden was behind the attacks.

"If it was bin Laden, then America will want revenge," I said.

Saima nodded. "And that will be dangerous for everybody, everybody in Afghanistan. America can't wipe him out in just one day," she said.

"It will be a family war," I said. "Bin Laden was for many years used by the CIA when they helped him fight the Russians, and now he is rebelling against America and the father is angry. But many of our people will pay the price for this."

"And afterward," Saima said, "what if America does punish the Taliban, and they lose power, what will happen then?"

None of us had an answer.

Later, RAWA told me to go ahead with my trip. After boarding, I sat in my seat thinking of the people who had been in the towers and of what they would have wanted to say to their parents or children if they had any understanding of what was happening to them. I was trying not to feel afraid of the flight ahead—I always feel a little claustrophobic in planes—when a Western couple came to sit next to me.

They were both middle-aged, dressed in jeans and T-shirts. We did not speak to one another, but from their accent I guessed that they were Americans.

"Those Afghans," the man said to his wife. "How can we let them do this to us?"

"Bush should just go ahead and bomb the place," she said.

I sat quietly, wondering how they would react if I told them I was from Afghanistan, that I had never protected bin Laden, and that I hated him as much as they did. A river red with blood separated the innocent people of Afghanistan from a handful of terrorists. I did not want to see bin Laden and the one-eyed Mullah Omar—the spiritual leader of the Taliban, who likes to describe himself as the *Amir-ul Momineen*, the Commander of the Faithful—killed right away. First I would take them, in a cage, around the most famous zoos of the world so that people could see what wild animals they were.

But I did not speak to them. I tried to concentrate instead on the film that was playing. Of all the films available, it was a comedy, *Mr. Bean*.

At Dubai airport, where I was due to change planes, I was stopped at passport control. The officer stared at the cover of the passport I had been given by the Taliban embassy. Islamic Emirate of Afghanistan, it said on the cover.

"Who are you visiting in Spain?" the officer asked me, turning the passport over in his hands again and again.

"My family. My aunt lives in Madrid," I lied.

214

My answer didn't convince him. "Please wait for me. I need to take your passport with me for a short while," he said.

Fifteen minutes later, he came back and handed my passport to me. "I'm sorry," he said. "The flight is canceled. There is nothing else available, and you will have to wait three days before we can tell you whether there is room on another flight. If you want to wait . . ."

It had been his turn to lie. "Why can't I fly? Is it the passport?" I insisted.

"Sorry. There's nothing I can do. We have received instructions to be very careful with Afghan nationals. That's all I can say," he replied.

Through no fault of my own, my passport and my nationality had become a liability. I had no choice but to call RAWA again and find a plane to return to Pakistan.

As I sat waiting for my flight, I spent hours watching CNN. I felt that the Afghans, because of all they had suffered, were the people who could best feel the pain of the people in New York and Washington.

I watched the archive footage of bin Laden. The way he dressed, the way he sat silently with downcast eyes or a lost, mystical expression—I thought he was trying to play the part of a prophet.

When I got back to Islamabad, I found our computer operator, Mehmooda, sadder than I had seen her for a long time. It is she who spends hours at a time dealing with the E-mails we receive from people all over the world. She drinks coffee to keep herself going through the long nights when she works because that is when the telephone rates are lowest.

"Most of the E-mails we've received are supportive, but you can't imagine how much hate mail we've had," Mehmooda told me. She showed me some of the messages. They were full of insults, but worst of all, many were from people who had previously been among our supporters. They said they did not want to raise funds for RAWA anymore because they hated Afghanistan. Up to ten percent of the E-mails we had received since September 11 were negative, many of them obscene.

One man wrote, "Someday soon people will say: 'You know, there used to be mountains in Afghanistan.'" Another, who signed himself as Lee, wrote, "Get out of your stupid country while you can. We (the USA) are going to blow you up with a nuclear bomb. You people should get rid of that stupid cloth on your head and join the real world. You rag-heads."

"What should I do?" Mehmooda asked me.

"Answer them. We should write an answer. Let's make it as kind as possible and send it to as many people as we can," I suggested.

Mehmooda wrote a message that said we understood their anger at such a moment, that we too were shocked by the attacks and shared the anger and sorrow of the American people. We pointed out that we too were victims of the Taliban, whom we called "a handful of brutal subhumans," and other Moslem fundamentalists in Afghanistan.

Many people who received this message E-mailed us back, apologizing for their earlier message.

"ANOTHER BOMB has just landed not very far away, but don't worry, we are all fine."

The voice of Shabnam was faint, the line to Kabul crackly. It had taken me an hour and a half to get through to her on the telephone from Peshawar. Shabnam, one of our RAWA friends in the capital, told me that already several civilians had been killed, promised to send us pictures of the damage done by bombings as soon as possible, and then the line went dead.

The previous night I had sat up until two o'clock in the morning watching the lights flash across the television screen as the Americans launched their offensive. I stared at the screen, but in my mind's eye I could see each of my friends in Kabul as clearly as if they were standing before me.

In the house where I was living at the time, CNN had been on from early morning to the small hours of the night, every day since the September 11 attacks, as we waited and waited.

Later, a member who managed to travel from Kabul to the frontier and then into Pakistan told me that right at the beginning of the offensive, a plane mistook a truck distributing water for an oil tanker and dropped a bomb on it. There was a severe drought, and the truck was in a central neighborhood of Kabul. The bomb destroyed five houses.

It was impossible for Washington to strike only at bin Laden. Many innocent people would die first.

THE MOTHER BLAMED HERSELF, although it was not her fault. She kept repeating that she and her husband had carried her six-year-old son for miles at night over the mountains—they only walked at night, to avoid the Taliban patrols—but that when she felt too tired and too weak to carry him any further, she had set him down on the narrow path and made him walk ahead of her. The path ran along the edge of a precipice, and in the darkness she was just able to see him slip and vanish down the mountainside.

"I was selfish, and this is my punishment," said the woman, who was from the Uzbek tribe, as she sat on the floor of my house in the refugee camp. She had fled to Pakistan with her husband and their child when Mullah Omar declared a jihad, or holy war, against America and ordered that every family must designate one man who would enlist to fight. Pakistan had promptly closed its borders with Afghanistan to stop Afghans from escaping.

There was no room for the couple in the camp because so many refugees had arrived in the last few weeks, so Ameena invited the woman to stay in our house until we found a solution. Her plastic shoes were in bits, held together by string, and her feet were bloody from her long walk. She had spent all she had on her journey. She asked us to help her find work making carpets, and for her husband, work in the brick factory.

She had wanted to stay in the mountains and look for her son's body, but her husband had said it was too dangerous and they had to move on. He told her there was no chance that the boy could have survived such a fall.

Chapter Nineteen

FOR DAYS we had heard shouts, scuffling and banging sounds from the room in the school where the girls were rehearsing. "You are all infidels!" was one cry that I often heard when I walked past. I was preparing for the coming visit to the refugee camp of a delegation of European parliamentarians. Sometimes I thought the girls were busier than Ameena and I were.

Neither Ameena nor I was allowed to enter the room. Little Shamms was always warning me to stay away. He had managed to get himself mixed up in the girls' preparations for the play that they would put on for the visitors. The girls he knew as his sisters tried to give him a part in the play, but he was always forgetting his lines—he would stop short and ask, "What do I say now?"—so they gave up and used him as an assistant instead.

The girls were aged between seven and fourteen. They were from very poor families, but they were not short of ideas. Day after day, often several times a day, they sent us Shamms as their messenger to request something for the play.

After he had knocked at my door a dozen times—he always knew when I was in because he could spot my shoes outside the door—I pulled Shamms's leg: "Tell me the truth," I said. "You are the one responsible for all this, aren't you?"

He looked pleased, but he shook his head. "No, no, the older children are," he said.

"I am sure that without you, this play would not be possible," I said.

Shamms ran away, fighting hard to hide a big grin.

They borrowed turbans from traditional families, boys' clothes, a whip that we had made out of plastic and rope for a previous play. For each request, I had to write a note for Shamms to deliver. He didn't know how to read, but he quickly realized the power that these notes gave him, and he would fold them up twice, very neatly, and place them in the breast pocket of his shirt. He loved feeling like a Very Important Person.

"The older children" even wanted a dozen Kalashnikovs. I wrote to one of the security guards at the camp to lend just one to the children—and would he please make sure there were no bullets in it. One Kalashnikov would have to be enough. It would have symbolic value.

"Are you happy, Shamms?" I asked him one day. "Is everyone treating you with respect?"

"Yes, they are," he answered, puffing out his little chest.

Shamms informed me that now "the older children" needed me to summon a male supporter who could play the electronic organ to accompany them when they sang. The man was busy with other commitments, but I eventually gave in and asked him to come.

On the day the delegation arrived, the children were in a frenzy. As soon as Ameena and I sat down with the visitors to talk to them about the camp, Shamms knocked on the door and sidled up to me.

"Auntie, when are they coming to our school?" he asked me in a whisper.

I shooed him away, telling him sternly that if he bothered us again I would report him to his teacher.

Late that afternoon, there was hardly an hour left before the visitors would have to leave, and I was still showing them around. They couldn't stay later, as it was getting dark, and Peshawar was no place to travel around at night. There were many other things I wanted to show them in the camp, but I decided to take them to see the play because the children would have been immensely disappointed if I had failed to do so.

When we entered the schoolyard, we were greeted by two rows of girls who showered us with petals of flowers as a sign of greeting. We all took our seats—on the small chairs taken from the classrooms—in front of the stage, which was made of dried mud. When it rained heavily, the stage had to be rebuilt.

Shamms was running around looking very busy. His clothes had been washed and ironed. He had taken a shower and had also washed his hair. I saw him carefully running a comb through his fringe and grabbed him "You'll be needing some perfume next," I teased. "You are becoming like the girls. Don't forget that you are a boy." He grimaced and twisted himself free.

I went up to the stage and tried to get behind the curtain. I caught a glimpse of some beards and some veils before the girls stopped me, saying I must not see anything.

I promised not to look but demanded that they tell me what they were going to show us. Despite their protests, I had to cancel several of the sketches, songs, and poems they had prepared because there was so little time.

After what seemed to me to be a very long wait, a hand appeared to pull the curtain open, and the show began. A teenager gave a short speech thanking the foreign guests for coming to the camp. Again, another long wait as some bumping sounds came from behind the curtain.

I suddenly realized that I had no idea what we were about to see. Surely I should have insisted on seeing a rehearsal, I thought.

The curtain parted to reveal Osama bin Laden—or rather, a girl who was dressed in the white robes of a Saudi Arabian, with a beard cut out of a black rubbish bag stuck to her cheeks. Bin Laden was silent and spent his time staring down at the stage. He was surrounded by fawning Taliban guards in their turbans, also wearing beards. The girls' clothes were deliberately dirtied, to make them as filthy as the ones the Taliban wore. One of the guards carried the Kalashnikov, and I could see that the girl was afraid of it.

As another girl translated for the visitors, the Taliban burst in on a class in which the children were learning English.

"What is this?" the Taliban exclaimed in deep voices, holding a book upside down to show they were illiterate. "What are you being taught? Do you want to become infidels and prostitutes?"

The Taliban arrested the children, their parents, and the teacher and dragged them before bin Laden. "Cut their heads off," he commanded, without looking up.

But the teacher, together with the children, fought back against the Taliban, pulling at their beards, kicking and slapping them. Bags of red ink exploded on the stomachs of the Taliban, but the beatings looked as if they were for real.

When the Taliban realized they were being overrun, they turned against bin Laden, shouting at him, "Get out of our country!" He fled from the stage, humiliated, his robe flapping around his legs.

After a few songs and poems that were against both the Taliban and their fundamentalist enemies in the Northern Alliance, the curtain closed again to loud applause from us all. I felt proud of the children because they had done this on their own. Their play was childish, but it was a powerful message of resistance and of defiance against a terrorist whom the West had labeled the world's most wanted man. Even the Taliban, if defeated, would rid themselves of bin Laden as fast as they could, the children were saying.

For days afterward, the girls had strips of black stuck to their cheeks. The glue they had used was so strong that it was some time before they looked normal again.

NO ONE in the refugee camp was sorry to see the Taliban defeated. But no one rejoiced when the Taliban fled Kabul and the fighters of the Northern Alliance, which was made up of several veteran Mujahideen groups, took over the capital. Across the camp, people said, "Condolences," to one another and shook their heads, as if someone had died in the family. We all knew that although they now spoke of democracy, elections, and even women's rights, the Northern Alliance leaders who had taken power had blood on their hands. They were the same warlords who had bombed their own people in the early 1990s.

I spoke to three widows who live together in the camp. All three of them had lost their husbands to the Mujahideen. "What can we do now?" they asked me. "We have lost all the hope we had that one day we would be able to go back home." They had been counting on the return of the exiled king, Mohammed Zahir Shah, once the Taliban were defeated.

No one told me they wanted to return to Afghanistan. Some men of the Hazara tribe—the tribe Shamms belongs to—told me that if they returned to Kabul now, their eyes would be gouged out by the men of the Northern Alliance, or their necks would be cut with the technique of "the Dance of the Dead Body." Some refugees said that now the world was watching Afghanistan, perhaps these crimes would not be repeated.

In the days that followed, more refugees arrived at the camp. They had endured the American offensive, but now that the warlords had returned, they had fled because they had not forgotten the crimes committed only a few years earlier and feared that their daughters would be raped.

Whatever their promises, I do not believe that the Northern Alliance will bring peace and democracy to my country. The only goal of each faction is to have power for itself, and none of them are ready to share it. A civil war is the most likely outcome. Only a United Nations force could end the wars in my country by disarming all the warlords and overseeing free elections. And only a democratic and secular government could guarantee human rights, including women's rights.

THERE IS a very old song of Afghan folklore that I have always liked. It is in Pashto, and the refrain says, "I am ready to die for my love, but I want my love to be ready to die for my country."

I was stunned when I learned that Farah, who had studied at the school with me and then joined RAWA, was to marry a boy who had been chosen for her by her father. When Farah told me she was engaged, I could not believe it. The boy was studying in the *madrassa* Islamic school, and she barely knew him.

"How can you do this?" I asked her. "Do you know what it means, the fact that he studies in the *madrassa*?"

"If he goes to the *madrassa*, it's because his parents want him to," Farah said simply.

I thought that love—and I doubted this was real love—was not enough. "How can you be sure that this boy will respect you, if he is being taught in such a school? Have you forgotten all the ideas you were taught, and that you tried to spread among our people? I never expected this of you," I said.

She was silent.

There could be no greater gap between two different worlds than that between the boy's school and the school where Farah had been taught about women's rights and freedom throughout her teenage years. At the very least she knew that she should be the one to choose her husband, not her father.

Farah was betraying her principles. Perhaps she had grown tired of our work. Perhaps she was no longer prepared to risk her life, and she was seeking a normal life instead with a home, a husband, and children.

I do not want to give birth to a child. If there is one thing I have learned from my own life and from the refugee camp, it is that you can love a child even if it does not come from your womb. It is not important that the child is of your own blood. What is important is that you bring it up properly and love it.

There are too many orphaned or abandoned children whose only home is the street. Perhaps I could adopt one of them.

I HAVE NEVER SEEN Shamms cry. He still doesn't know that his parents are dead, and he never asks us what happened to them. But a couple of the teachers have seen him cry a few times as he lies in his bed. They ask him why he is crying, but he never tells them the reason. We dread him asking about his parents and hope he will not ask for a while yet.

One day we will tell Shamms the truth about them and about his fourteen sisters. It may be left to me to do this, when he is about ten years old. I have no idea how I will tell him. Perhaps I will say that many people have lost their parents in the war, and that although we cannot have our parents back, we can have our country back. Perhaps I will tell him about my own parents, about the example that they gave me. But I will never forget that I was lucky enough to spend much more time with my parents than he did with his, that they taught me much more than his parents ever could, and that I have always had Grandmother.

I keep Mother's white nightshirt in my room. Sometimes I take it out and press it to my face. I have never forgiven the men responsible for the deaths of my parents and of so many of my people. I cannot even begin to imagine forgiving them. If these men were to be taken before a court, I would like to see them punished not only for the deaths of my parents but also for all the crimes they have committed against my country.

When I think of my parents, I think of what they wanted me to accomplish. They did not want a daughter who would think only about herself. Sometimes, but not often, I wish that I could show them what I am doing. I know that I have not done enough in my life so far, and sometimes this saddens me, and I hope I will achieve something in the future.

But I am grateful that I was never alone. Grandmother was there when I lost my parents, and now she has come to live in the refugee camp where I sometimes work to be closer to me. "I didn't want to stay in an empty house all on my own," she said to me when she moved to the camp.

She is now in her seventies. She still reads the Koran and prays, using the same beads she had when I was a child. But because of the pain in her back and in her feet, she is no longer strong enough to stand up during the prayer, so she simply rocks backward and forward, touching the carpet with her forehead.

When I am away for a few weeks, she gets very little sleep. She is angry with me when I return, and tries not to speak to me to show her anger. She is very weak and very tired. She cries a little and says that if she cannot see me more often, she might as well die. She would like to see me every week.

I apologize to her. I tell her that I love her and that I cannot stop my work. Soon she breaks into a smile and asks Allah to forgive me for what I put her through. She pulls me over so that I can rest my head in her lap, and massages my head. I still have the little red knife she gave me when we celebrated my birthday in the shelter. When I can, I bring her some perfume.

I cannot imagine living any other kind of life than the one I have chosen. When RAWA sends me to Kabul or anywhere else, I do not go because I have to, but because I believe in what I am doing. I thought a great deal before choosing this kind of life, and I will not go back on that decision.

I have learned to live with fear. When you believe that danger is always present, you no longer feel the fear.

When I travel to the West, I never forget that my friends are in danger. I am always so worried, so sad, and feel under so much pressure that I never really enjoy the places I go to. I know that they are beautiful. When I was a child, I read about ancient Rome in the history books, and I remember seeing pictures of the Colosseum. I have been to Italy six times, but I have never gone inside the Colosseum. I would love to go, but not now.

Before I left Afghanistan, I thought my future was very dark, that there was no hope of a better life for me or for my country. I thought that my people were exhausted after suffering war for so many years. They had thrown out the Russians, but they no longer had the strength to rise up against the fundamentalists. But the school I went to gave me hope. It taught me that education and respect for the rights of both men and women could change society. I am a little over twenty years old, and my greatest desire is that peace returns to my country. I wonder whether, after more than twenty years of fighting in Afghanistan, the world has understood the real nature of fundamentalism, whether I will again hear the sounds of foreign soldiers marching into my country, of the Kalashnikov, of people crying. But I know that I will never lose hope and that I will continue to battle for the ideals I believe in, the ideals for which Meena, the founder of RAWA, sacrificed her life.

If peace returns to my country, I would like to go back and walk the destroyed streets of Kabul, the sun shining not on a *burqa* but on my face. I would think not of the past but of the future. I would show Shamms the streets of my childhood, take him into my house, and teach him how to fly a kite from our roof. And I would tease him when the kite slipped from his grasp and flew away on its own, higher than the mountains.

Postscript

WE MET ZOYA for the first time at the small hotel where she was staying near the Vatican during a fund-raising trip to Italy in the spring of 2001. Rita had read about her sometime before and tracked her down to Pakistan with the help of Amnesty International and Emergency, an Italian relief organization that helps war victims. We had then won permission to interview Zoya for the two magazines we work for.

She joined us in the hotel lobby. She was friendly and attractive and wore a gray dress with a scarf draped over her shoulders. We knew that in her homeland not even that scarf would have saved her from a beating in the street.

Her gaze intense, she spoke with a confidence that made her seem older than her twenty-three years. Her graceful manner was at odds with the stories she told us, stories that often seemed inconceivable to us. But we felt close to her, and we identified with her hopes. It was her optimism that struck us: she wanted to teach women to read and write in a country where most of them were illiterate, to treat sick women in a country where the authorities decreed they should die rather than be treated by male doctors, to speak of justice and democracy in a country where the only law was that of an eye for an eye, a tooth for a tooth.

After the interview, we asked her if we could go for a walk together. We felt the need to spend some more time with her, without a tape recorder. It was raining, but she quickly accepted. She didn't care about the rain—she didn't even want to use an umbrella. She told us that she had been exiled to a desert and enjoyed the rain. When we got back to the hotel, she asked us to wait and went to fetch something from her room. It was a present, a hand-carved iron box set with black stones, and inside were a necklace, a bracelet, a ring, and earrings made of silver and blue enamel. Zoya told us they had all been made in a refugee camp.

When we telephoned her sometime later to offer to write a book with her, her first questions were "Why don't you write a book about someone else? What is special about my story?" When she finally agreed, she said she wanted the book to stand for the suffering of all Afghan women. She came to stay with us in Rome so that we could ask her at length about her life. Only the people closest to her knew where she was, for her own protection.

As we listened, she led us into her world. We felt, with her, the claustrophobia of the *burqa*, we heard the Taliban's whip whistling through the air, and we saw the tears of the mothers who had lost their sons. But Zoya also made us laugh with her contagious sense of humor about the absurd aspects of life under the rule of religious fundamentalists.

We have several people to thank for helping to make this book happen: Cristina Cattafesta at Emergency and Edoardo Bai; Luca Lopresti at Amnesty International; Sean Ryan, Paolo Palleschi, and Camillo Ricci, our employers, who gave us the time to write the book; Anne Keefe, who transcribed the tapes of our interviews at speed and cheered us on; our publisher, Michael Morrison, who backed the idea with enthusiasm; our editor, Claire Wachtel, who flew to join us for the last sprint; our agent, Clare Alexander, who guided us from the beginning; and our families, who warmly encouraged us.

We hope that sooner or later we will visit Zoya in Kabul and that we will find her living the life she wants to lead.

JOHN FOLLAIN AND RITA CRISTOFARI
Rome, January 2002

Chronology

1839-42 First Anglo-Afghan war.

1878-80 Second Anglo-Afghan war. Britain wins
control of Afghanistan's foreign affairs.

1880-1901 Amir Abdul Rahman Khan conquers
Afghanistan. Current borders of
Afghanistan are determined.

1933-73 King Mohammed Zahir Shah rules.

1959 Prime Minister Daoud and other
government ministers appear at national
celebrations with their wives and daughters
unveiled, sparking riots led by religious
leaders.

1964 Constitution proclaiming legal equality of men and women prompts more unrest.

July 1973 King Zahir Shah is overthrown by Daoud while on holiday in the Bay of Naples.

April 1978 Marxist-Leninist military coup. People's Democratic Party of Afghanistan takes power.

December 1979 Soviet troops invade. Mujahideen fighters launch battle against the occupiers.

February 1986 Soviet leader Mikhail Gorbachev describes Afghanistan as a "bleeding wound," indicates that he wants to pull troops out.

February 1989 Soviet troops withdraw.

April 1992 Mujahideen government takes power in Kabul. President Najibullah seeks refuge in UN compound.

1993 Some ten thousand civilians die in battles between President Rabbani and Gulbuddin Hikmetyar, a fundamentalist Mujahid leader.

November 1994 Taliban conquer Kandahar.

September 1995 Taliban conquer Herat.

November 1995 Taliban bomb Kabul before government troops force them back.

September 1996	Taliban conquer Jalalabad and Kabul. Taliban hang Najibullah and his brother.
August 1998	Osama bin Laden blamed for bombings of U.S. embassies in Kenya and Tanzania. Mullah Omar pledges to give protection to bin Laden. U.S. cruise missile attack on bin Laden camps in Jalalabad and Khost.
June 1999	FBI places bin Laden at top of list of ten most wanted outlaws.
January 2001	Taliban massacre about three hundred civilians in Yakaolang.
March 2001	Taliban destroy giant Buddha statues in Bamiyan Valley.
September 2001	Bin Laden blamed for airplane attacks on the World Trade Center and the Pentagon.
October 2001	United States and Britain lead bombing offensive against Al'Qaida, bin Laden's organization, in Afghanistan.
November 2001	Northern Alliance conquers Kabul.